Synthesis Lectures on Mathematics & Statistics

Series Editor

Steven G. Krantz, Department of Mathematics, Washington University, Saint Louis, MO, USA

This series includes titles in applied mathematics and statistics for cross-disciplinary STEM professionals, educators, researchers, and students. The series focuses on new and traditional techniques to develop mathematical knowledge and skills, an understanding of core mathematical reasoning, and the ability to utilize data in specific applications.

Can Chen

Tensor-Based Dynamical Systems

Theory and Applications

Can Chen
School of Data Science and Society
and Department of Mathematics
University of North Carolina at Chapel Hill
Chapel Hill, NC, USA

ISSN 1938-1743   ISSN 1938-1751   (electronic)
Synthesis Lectures on Mathematics & Statistics
ISBN 978-3-031-54507-8   ISBN 978-3-031-54505-4   (eBook)
https://doi.org/10.1007/978-3-031-54505-4

This Springer imprint is published by the registered company Springer Nature Switzerland AG
The registered company address is: Gewerbestrasse 11, 6330 Cham, Switzerland

Paper in this product is recyclable.

To my family

Preface

Tensors are multidimensional arrays that generalize vectors and matrices, enabling the representation of intricate higher-order interactions within multiway data. Tensor-based dynamical systems extend the traditional notion of linear dynamical systems to a more general setting where system evolutions are governed by tensor products, allowing us to model and analyze a wider range of real-world applications with greater precision and efficiency.

Tensor-based dynamical systems have garnered significant attention due to their ability to leverage tensor algebra to accelerate system-related computations and to facilitate deeper understandings of higher-order dynamical interactions. Furthermore, they provide a unified and flexible approach to modeling and analyzing various types of systems, encompassing physical systems, chemical reactions, biological processes, social networks, and more. In recent years, there has been a growing interest in developing efficient computational methods for analyzing and controlling different types of tensor-based dynamical systems.

Overall, tensor-based dynamical systems have emerged as a groundbreaking mathematical framework with the potential to revolutionize our understanding and control of complex systems. Consequently, research in this field holds immense significance across diverse disciplines, including mathematics, computer science, engineering, biology, social science, and beyond.

Chapel Hill, NC, USA   Can Chen

Acknowledgements

I am indebted to Dr. Anthony M. Bloch and Dr. Indika Rajapakse at the University of Michigan for their visionary introduction to the field of tensor-based dynamical systems and their unwavering support during my doctoral studies. My early explorations in the field were also enriched by the scientific guidance of Dr. Amit Surana at Raytheon Technologies Research Center, to whom I extend my sincere gratitude. I am also deeply grateful to Dr. Yang-Yu Liu at Harvard Medical School for his encouragement and support throughout the writing of this book.

Contents

Acronyms

AIR   Aachen Impulse Response
BPOD   Balanced Proper Orthogonal Decomposition
BT   Balanced Truncation
CPD   CANDECOMP/PARAFAC Decomposition
CPDS   Contracted Product-based Dynamical System
EM   Expectation Maximization
EPDS   Einstein Product-based Dynamical System
ERA   Eigensystem Realization Algorithm
HOSVD   Higher-order Singular Value Decomposition
HTD   Hierarchical Tucker Decomposition
HVAC   Heating, Ventilation, and Air Conditioning
LDS   Linear Dynamical System
MIMO   Multiple Inputs Multiple Outputs
PCA   Principal Component Analysis
SISO   Single Input Single Output
SVD   Singular Value Decomposition
TD   Tucker Decomposition
t-PDS   t-Product-based Dynamical System
TPDS   Tucker Product-based Dynamical System
TSVD   Tensor Singular Value Decomposition
t-SVD   t-Singular Value Decomposition
TTD   Tensor Train Decomposition
TVPDS   Tensor Vector Product-based Dynamical System

Tensor Preliminaries

1

Abstract

Tensors are multidimensional arrays generalized from vectors and matrices, which have a broad range of applications in various fields such as signal processing, machine learning, statistics, dynamical systems, numerical linear algebra, computer vision, neuroscience, network science, and elsewhere. Given the widespread use of tensors, understanding tensor algebra is therefore essential when working with them. Tensor algebra encompasses a wide range of topics as linear algebra, including tensor products, tensor unfoldings, block tensors, tensor eigenvalues, and tensor decompositions, each of which plays a critical role in diverse applications.

1.1 Overview

The word tensor derives from the Latin *tendere*, meaning "to stretch." It first emerged with its modern physical sense from the work of Voigt [1] in 1898 to describe stress and strain on crystals. As a matter of fact, the concept of tensors had already been employed in the context of continuum mechanics [2–4] and differential geometry [5–7]. Following the development of Einstein's theory of general relativity [8, 9], tensors gained considerable attention within the physics and mathematics communities. Tensors can be understood in various prospectives with different levels of abstraction. This book considers tensors as multidimensional arrays generalized from vectors and matrices.

Tensors find widespread use in a variety of fields, including signal processing [10–12], machine learning [13–15], statistics [16–18], dynamical systems [19–22], numerical linear algebra [23–25], computer vision [26–28], neuroscience [29–31], network analysis [32–34], and elsewhere. Of particular interest of this book is exploring the role of tensor algebra in tensor-based dynamical systems, where the evolutions of the systems are captured by tensor

© The Author(s), under exclusive license to Springer Nature Switzerland AG 2024

C. Chen, *Tensor-Based Dynamical Systems*,
Synthesis Lectures on Mathematics & Statistics,
https://doi.org/10.1007/978-3-031-54505-4_1

products. The dynamics of human genome is an excellent example of tensor-based dynamical systems, where the 3D genome structure, function, and its relationship to phenotype can be represented by tensors [35, 36]. Tensor algebra can be leveraged to accelerate system-related computations and facilitate understandings of the dynamical processes in tensor-based dynamical systems, making it an exciting area of research.

The order of a tensor refers to the total number of its dimensions, with each dimension called a mode. A kth-order tensor is typically denoted by $\mathsf{T} \in \mathbb{R}^{n_1 \times n_2 \times \cdots \times n_k}$. Hence, scalars $s \in \mathbb{R}$ are zero-order tensors, vectors $\mathbf{v} \in \mathbb{R}^n$ are first-order tensors, and matrices $\mathbf{M} \in \mathbb{R}^{n \times m}$ are second-order tensors. Tensors can be divided into sub-arrays by fixing a subset of their indices. A fiber is a sub-array obtained by fixing all but one index, while a slice is obtained by fixing all but two indices. For example, a third-order tensor $\mathsf{T} \in \mathbb{R}^{n_1 \times n_2 \times n_3}$ can have fibers commonly named as columns, rows, and tubes, and its slices are commonly named as horizontals, laterals, and frontals, see Fig. 1.1. The colon notation ":" refers to the MATLAB colon operation, which acts as shorthand to include all subscripts in a particular array dimension.

If all modes of a kth-order tensor T have the same size, i.e., $n_1 = n_2 = \cdots = n_k$, it is often called a cubical tensor. A cubical tensor is called supersymmetric if $\mathsf{T}_{j_1 j_2 \cdots j_k}$ is invariant under any permutation of the indices. For example, a third-order tensor $\mathsf{T} \in \mathbb{R}^{n \times n \times n}$ is supersymmetric if it satisfies the following condition:

$$\mathsf{T}_{j_1 j_2 j_3} = \mathsf{T}_{j_1 j_3 j_2} = \mathsf{T}_{j_2 j_1 j_3} = \mathsf{T}_{j_2 j_3 j_1} = \mathsf{T}_{j_3 j_1 j_2} = \mathsf{T}_{j_3 j_2 j_1}.$$

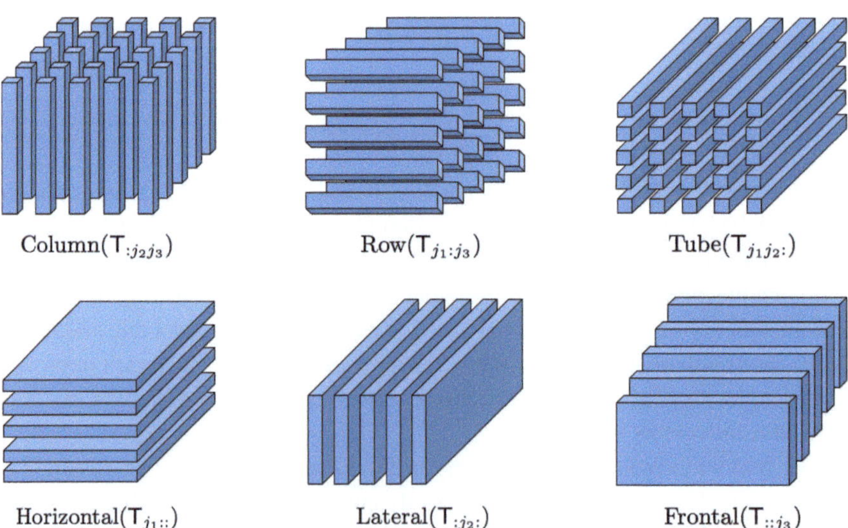

Fig. 1.1 Fibers and slices of a third-order tensor

A cubical tensor is called diagonal if $\mathsf{T}_{j_1 j_2 \cdots j_k} = 0$ except $j_1 = j_2 = \cdots = j_k$. Additionally, even-order tensors, often denoted by $\mathsf{T} \in \mathbb{R}^{n_1 \times m_1 \times \cdots \times n_k \times m_k}$, are tensors with even orders, i.e., $2k$, which possess various interesting properties in tensor algebra [19, 37, 38].

This chapter provides a comprehensive review of tensor algebra, which is fundamental for the theory of tensor-based dynamical systems. Most of the concepts and notations for tensor algebra were adapted from the work of [19, 39–47].

1.2 Tensor Products

This section presents different types of tensor products, which are crucial in constructing tensor-based dynamical systems and developing their systems theory.

1.2.1 Kronecker Product

Although the Kronecker product is defined on matrices, it plays an important role in tensor algebra [48, 49]. Given two matrices of arbitrary size $\mathbf{A} \in \mathbb{R}^{n \times m}$ and $\mathbf{B} \in \mathbb{R}^{s \times r}$, the Kronecker product of the two matrices, denoted by $\mathbf{A} \otimes \mathbf{B} \in \mathbb{R}^{ns \times mr}$, is defined as

$$\mathbf{A} \otimes \mathbf{B} = \begin{bmatrix} \mathbf{A}_{11}\mathbf{B} & \mathbf{A}_{12}\mathbf{B} & \cdots & \mathbf{A}_{1m}\mathbf{B} \\ \mathbf{A}_{21}\mathbf{B} & \mathbf{A}_{22}\mathbf{B} & \cdots & \mathbf{A}_{2m}\mathbf{B} \\ \vdots & \vdots & \ddots & \vdots \\ \mathbf{A}_{n1}\mathbf{B} & \mathbf{A}_{n2}\mathbf{B} & \cdots & \mathbf{A}_{nm}\mathbf{B} \end{bmatrix}, \tag{1.1}$$

where $\mathbf{A}_{ji}$ denotes the (j, i)th entries of $\mathbf{A}$. The Kronecker product possesses many useful properties, including bilinearity, associativity, and the mixed-product property [49, 50]. The Kronecker product is also generalized to tensors, see details in [51].

1.2.2 Khatri-Rao Product

The Khatri-Rao product, like the Kronecker product, is defined on matrices, but it is important in tensor computation [49, 52]. Given two matrices $\mathbf{A} \in \mathbb{R}^{n \times s}$ and $\mathbf{B} \in \mathbb{R}^{m \times s}$, the Khatri-Rao product of the two matrices, denoted by $\mathbf{A} \odot \mathbf{B} \in \mathbb{R}^{nm \times s}$, is defined as

$$\mathbf{A} \odot \mathbf{B} = \begin{bmatrix} \mathbf{a}_1 \otimes \mathbf{b}_1 & \mathbf{a}_2 \otimes \mathbf{b}_2 & \cdots & \mathbf{a}_s \otimes \mathbf{b}_s \end{bmatrix}, \tag{1.2}$$

where $\mathbf{a}_j$ and $\mathbf{b}_j$ are the jth columns of $\mathbf{A}$ and $\mathbf{B}$, respectively. Readers may find more information on the properties of the Khatri-Rao product in [49].

1.2.3 Outer/Inner Product

The outer product between two tensors is a generalization of the vector outer product [39]. Given two tensors $\mathsf{T} \in \mathbb{R}^{n_1 \times n_2 \times \cdots \times n_{k_1}}$ and $\mathsf{S} \in \mathbb{R}^{m_1 \times m_2 \times \cdots \times m_{k_2}}$, their outer product, denoted by $\mathsf{T} \circ \mathsf{S} \in \mathbb{R}^{n_1 \times n_2 \times \cdots \times n_{k_1} \times m_1 \times m_2 \times \cdots \times m_{k_2}}$, is defined as

$$(\mathsf{T} \circ \mathsf{S})_{j_1 j_2 \cdots j_{k_1} i_1 i_2 \cdots i_{k_2}} = \mathsf{T}_{j_1 j_2 \cdots j_{k_1}} \mathsf{S}_{i_1 i_2 \cdots i_{k_2}}. \tag{1.3}$$

In contrast, the inner product between two tensors of the same size $\mathsf{T}, \mathsf{S} \in \mathbb{R}^{n_1 \times n_2 \times \cdots \times n_k}$, denoted by $\langle \mathsf{T}, \mathsf{S} \rangle \in \mathbb{R}$, is defined as

$$\langle \mathsf{T}, \mathsf{S} \rangle = \sum_{j_1=1}^{n_1} \sum_{j_2=1}^{n_2} \cdots \sum_{j_k=1}^{n_k} \mathsf{T}_{j_1 j_2 \cdots j_k} \mathsf{S}_{j_1 j_2 \cdots j_k}. \tag{1.4}$$

Two tensors T and S are said to be orthogonal if $\langle \mathsf{T}, \mathsf{S} \rangle = 0$. The tensor Frobenius norm is therefore defined as $\|\mathsf{T}\|^2 = \langle \mathsf{T}, \mathsf{T} \rangle$.

1.2.4 Tensor Matrix/Vector Product

The tensor matrix/vector product is an extension of familiar matrix matrix/vector products [39]. Given a kth-order tensor $\mathsf{T} \in \mathbb{R}^{n_1 \times n_2 \times \cdots \times n_k}$ and a matrix $\mathbf{M} \in \mathbb{R}^{m \times n_p}$, the tensor matrix product $\mathsf{T} \times_p \mathbf{M} \in \mathbb{R}^{n_1 \times n_2 \times \cdots \times n_{p-1} \times m \times n_{p+1} \times \cdots \times n_k}$ along mode p is defined as

$$(\mathsf{T} \times_p \mathbf{M})_{j_1 j_2 \cdots j_{p-1} i j_{p+1} j_{p+2} \cdots j_k} = \sum_{j_p=1}^{n_p} \mathsf{T}_{j_1 j_2 \cdots j_p \cdots j_k} \mathbf{M}_{i j_p}. \tag{1.5}$$

This product can be extended to the so-called Tucker product as follows:

$$\mathbf{T} \times_1 \mathbf{M}_1 \times_2 \mathbf{M}_2 \times_3 \cdots \times_k \mathbf{M}_k = \mathsf{T} \times \{\mathbf{M}_1, \mathbf{M}_2, \ldots, \mathbf{M}_k\} \in \mathbb{R}^{m_1 \times m_2 \times \cdots \times m_k}$$

for $\mathbf{M}_p \in \mathbb{R}^{m_p \times n_p}$. When $m = 1$, the tensor matrix product (1.5) reduces to tensor vector products. Given a vector $\mathbf{v} \in \mathbb{R}^{n_p}$, the tensor vector product $\mathsf{T} \times_p \mathbf{v} \in \mathbb{R}^{n_1 \times n_2 \times \cdots \times n_{p-1} \times n_{p+1} \times \cdots \times n_k}$ along mode p is defined as

$$(\mathsf{T} \times_p \mathbf{v})_{j_1 j_2 \cdots j_{p-1} j_{p+1} j_{p+2} \cdots j_k} = \sum_{j_p=1}^{n_p} \mathsf{T}_{j_1 j_2 \cdots j_p \cdots j_k} \mathbf{v}_{j_p}. \tag{1.6}$$

Similarly, it can be generalized to the Tucker product form, i.e., for $\mathbf{v}_p \in \mathbb{R}^{n_p}$,

$$\mathsf{T} \times_1 \mathbf{v}_1 \times_2 \mathbf{v}_2 \times_3 \cdots \times_k \mathbf{v}_k = \mathsf{T} \times \{\mathbf{v}_1, \mathbf{v}_2, \ldots, \mathbf{v}_k\} \in \mathbb{R}.$$

Suppose that $\mathsf{T} \in \mathbb{R}^{n \times n \times \overset{k}{\cdots} \times n}$ is supersymmetric and $\mathbf{x} \in \mathbb{R}^n$ contains k variables $x_1, x_2, \ldots, x_k$. The homogeneous polynomial of degree k associated with T can be expressed as

$$h(x_1, x_2, \ldots, x_k) = \mathsf{T} \times \{\mathbf{x}, \mathbf{x}, \overset{k}{\cdots}, \mathbf{x}\} = \mathsf{T}\mathbf{x}^k. \tag{1.7}$$

Previous research [21, 53, 54] has shown that every n-dimensional homogeneous polynomial of degree k can be uniquely determined by a kth-order n-dimensional supersymmetric tensor.

1.2.5   Einstein Product

The Einstein product is a tensor contraction defined over the space of even-order tensors (though it can be defined more generally for arbitrary-sized tensors) [19, 35]. Given two even-order tensors $\mathsf{T} \in \mathbb{R}^{n_1 \times m_1 \times \cdots \times n_k \times m_k}$ and $\mathsf{S} \in \mathbb{R}^{m_1 \times s_1 \times \cdots \times m_k \times s_k}$, the Einstein product of the two tensors, denoted by $\mathsf{T} * \mathsf{S} \in \mathbb{R}^{n_1 \times s_1 \times \cdots \times n_k \times s_k}$, is defined as

$$(\mathsf{T} * \mathsf{S})_{j_1 l_1 \cdots j_k l_k} = \sum_{i_1=1}^{m_1} \sum_{i_2=1}^{m_2} \cdots \sum_{i_k=1}^{m_k} \mathsf{T}_{j_1 i_1 \cdots j_k i_k} \mathsf{S}_{i_1 l_1 \cdots i_k l_k}. \tag{1.8}$$

The Einstein product can be viewed as the multidimensional extension of matrix matrix products. Therefore, it can also be defined between a $2k$th-order tensor and a kth-order tensor, similar to matrix vector products. Given $\mathsf{T} \in \mathbb{R}^{n_1 \times m_1 \times \cdots \times n_k \times m_k}$ and $\mathsf{X} \in \mathbb{R}^{m_1 \times m_2 \times \cdots \times m_k}$, the Einstein product $\mathsf{T} * \mathsf{X} \in \mathbb{R}^{n_1 \times n_2 \times \cdots \times n_k}$ is defined as

$$(\mathsf{T} * \mathsf{X})_{j_1 j_2 \cdots j_k} = \sum_{i_1=1}^{m_1} \sum_{i_2=1}^{m_2} \cdots \sum_{i_k=1}^{m_k} \mathsf{T}_{j_1 i_1 \cdots j_k i_k} \mathsf{X}_{i_1 i_2 \cdots i_k}. \tag{1.9}$$

The space of even-order tensors with the Einstein product possesses many desirable properties for tensor computation [19, 41, 55]. For instance, the Tucker product can be efficiently represented in the Einstein product form for even-order tensors [19]. Given an even-order tensor $\mathsf{T} \in \mathbb{R}^{n_1 \times m_1 \times \cdots \times n_k \times m_k}$ and matrices $\mathbf{U}_p \in \mathbb{R}^{s_p \times n_p}$ and $\mathbf{V}_p \in \mathbb{R}^{r_p \times m_p}$ for $p = 1, 2, \ldots, k$, the Tucker product can be rewritten as

$$\mathsf{T} \times \{\mathbf{U}_1, \mathbf{V}_1, \ldots, \mathbf{U}_k, \mathbf{V}_k\} = \mathsf{U} * \mathsf{T} * \mathsf{V}^\top \in \mathbb{R}^{s_1 \times r_1 \times \cdots \times s_k \times r_k}, \tag{1.10}$$

where $\mathsf{U} = \mathbf{U}_1 \circ \mathbf{U}_2 \circ \cdots \circ \mathbf{U}_k \in \mathbb{R}^{s_1 \times n_1 \times \cdots \times s_k \times n_k}$ and $\mathsf{V} = \mathbf{V}_1 \circ \mathbf{V}_2 \circ \cdots \circ \mathbf{V}_k \in \mathbb{R}^{r_1 \times m_1 \times \cdots \times r_k \times m_k}$. Here, the notation $\mathsf{V}^\top \in \mathbb{R}^{m_1 \times r_1 \times \cdots \times m_k \times r_k}$ is referred to as the U-transpose of V, which is defined as

$$\mathsf{V}^\top_{i_1 j_1 \cdots i_k j_k} = \mathsf{V}_{j_1 i_1 \cdots j_k i_k}.$$

Note that the superscript "$\top$" is used to represent the operations of matrix transpose, U-transpose, and t-transpose (defined below), so it is important to understand the context in which it is being used. Further interesting properties regarding the Einstein product in the space of even-order tensors will be discussed.

1.2.6 t-Product

The t-product is a powerful tool for manipulating third-order tensors, allowing for multiplication between them based on the concept of circular convolution [44, 46, 47]. Given two third-order tensors $T \in \mathbb{R}^{n \times m \times s}$ and $S \in \mathbb{R}^{m \times r \times s}$, the t-product $T \star S \in \mathbb{R}^{n \times r \times s}$ is defined as

$$T \star S = \mathtt{fold}\Big(\mathtt{bcirc}(T)\,\mathtt{unfold}(S)\Big), \qquad (1.11)$$

where

$$\mathtt{bcirc}(T) = \begin{bmatrix} T_{::1} & T_{::s} & \cdots & T_{::2} \\ T_{::2} & T_{::1} & \cdots & T_{::3} \\ \vdots & \vdots & \ddots & \vdots \\ T_{::s} & T_{::(s-1)} & \cdots & T_{::1} \end{bmatrix} \in \mathbb{R}^{ns \times ms}, \quad \mathtt{unfold}(S) = \begin{bmatrix} S_{::1} \\ S_{::2} \\ \vdots \\ S_{::s} \end{bmatrix} \in \mathbb{R}^{ms \times r},$$

and $\mathtt{fold}$ is the reverse operation of $\mathtt{unfold}$. The t-product of two tensors usually does not commute except for the case when $n = m = r = 1$. The t-inverse of a third-order tensor $T \in \mathbb{R}^{n \times n \times s}$, denoted by $T^{-1} \in \mathbb{R}^{n \times n \times s}$, is defined as $T \star T^{-1} = T^{-1} \star T = I$, where I is the t-identity tensor whose first frontal slice (i.e., $I_{::1}$) is the identity matrix and all other frontal slices are all zeros. The t-inverse of T can be computed as

$$T^{-1} = \mathtt{fold}\Big(\mathtt{bcirc}(T)^{-1}\Big).$$

Note that the superscript "-1" is used to represent both matrix inverse and t-inverse operations. The t-transpose of a third-order tensor $T \in \mathbb{R}^{n \times m \times s}$, denoted by $T^{\top} \in \mathbb{R}^{m \times n \times s}$, can be obtained by transposing each of the frontal slices and then reversing the order of transposed frontal slices 2 through s. Moreover, a third-order tensor $T \in \mathbb{R}^{n \times n \times s}$ is called t-orthogonal if $T \star T^{\top} = T^{\top} \star T = I$.

1.3   Tensor Unfoldings

Tensor unfolding is a fundamental operation in tensor algebra that maps a tensor to a matrix or vector representation [19, 39, 40]. For convenience, denote $\mathcal{J} = \{j_1, j_2, \ldots, j_k\}$ and $\mathcal{N} = \{n_1, n_2, \ldots, n_k\}$ as the sets of dimension indices and sizes of a kth-order tensor T, respectively. Let $\varphi(\cdot, \mathcal{N}) : \mathbb{Z}^+ \times \mathbb{Z}^+ \times \overset{k}{\cdots} \times \mathbb{Z}^+ \to \mathbb{Z}^+$ be an index mapping function defined as

$$\varphi(\mathcal{J}, \mathcal{N}) = j_1 + \sum_{p=2}^{k} (j_p - 1) \prod_{l=1}^{p-1} n_l. \tag{1.12}$$

The vectorization of T, denoted by $\mathrm{vec}(\mathsf{T}) \in \mathbb{R}^{\prod_{p=1}^{k} n_p}$, can be simply represented by utilizing the index mapping function, i.e.,

$$\mathrm{vec}(\mathsf{T})_j = \mathsf{T}_{j_1 j_2 \cdots j_k} \quad \text{for } j = \varphi(\mathcal{J}, \mathcal{N}).$$

1.3.1   Tensor Matricization

Tensor matricization can be formulated similarly by considering two index mapping functions. Let $z \in \mathbb{Z}^+$ such that $1 \leq z < k$ and $\mathbb{S}$ be a permutation of the set $\{1, 2, \ldots, k\}$. Define the sets of row and column indices as

$$\mathcal{I} = \{i_1, i_2, \ldots, i_z\} \quad \text{for } i_p = j_{\mathbb{S}(p)},$$
$$\mathcal{L} = \{l_1, l_2, \ldots, l_{k-z}\} \quad \text{for } l_p = j_{\mathbb{S}(z+p)},$$

with the corresponding size sets

$$\mathcal{M} = \{m_1, m_2, \ldots, m_z\} \quad \text{for } m_p = n_{\mathbb{S}(p)},$$
$$\mathcal{S} = \{s_1, s_2, \ldots, s_{k-z}\} \quad \text{for } s_p = n_{\mathbb{S}(z+p)},$$

respectively. The z-unfolding of T under the permutation $\mathbb{S}$, denoted by $\mathbf{T}_{(z,\mathbb{S})} \in \mathbb{R}^{\prod_{p=1}^{z} m_p \times \prod_{p=1}^{k-z} s_p}$, is defined as

$$(\mathbf{T}_{(z,\mathbb{S})})_{il} = \mathsf{T}^{\mathbb{S}}_{i_1 i_2 \cdots i_z l_1 l_2 \cdots l_{k-z}}, \tag{1.13}$$

where $i = \varphi(\mathcal{I}, \mathcal{M})$ and $l = \varphi(\mathcal{L}, \mathcal{S})$. Here, the notation $\mathsf{T}^{\mathbb{S}} \in \mathbb{R}^{n_{\mathbb{S}(1)} \times n_{\mathbb{S}(2)} \times \cdots \times n_{\mathbb{S}(k)}}$ is referred to as the $\mathbb{S}$-transpose of T, which is defined as

$$\mathsf{T}^{\mathbb{S}}_{j_{\mathbb{S}(1)} j_{\mathbb{S}(2)} \cdots j_{\mathbb{S}(k)}} = \mathsf{T}_{j_1 j_2 \cdots j_k}.$$

The `unfold` operation defined in the t-product can be represented using (1.13) with $z = 2$ and $\mathbb{S} = \{1, 3, 2\}$. Additionally, when $z = 1$ and $\mathbb{S} = (p, 1, 2, \ldots, p-1, p+1, p+2, \ldots, k)$, the unfolding is referred to as the p-mode matricization, and it is denoted by $\mathbf{T}_{(p)} \in \mathbb{R}^{n_p \times \prod_{q \neq p} n_q}$.

1.3.2 Isomorphic Unfolding

The isomorphic unfolding is a power theoretical construct for even-order tensors [19, 35, 41]. Given an even-order tensor $T \in \mathbb{R}^{n_1 \times m_1 \times \cdots \times n_k \times m_k}$, define a mapping ψ such that

$$T_{j_1 i_1 \cdots j_k i_k} \xrightarrow{\psi} \mathbf{T}_{ji}, \tag{1.14}$$

where $j = \varphi(\mathcal{J}, \mathcal{N})$ and $i = \varphi(\mathcal{I}, \mathcal{M})$. The mapping can also be viewed as a tensor matricization $\mathbf{T}_{(z, \mathbb{S})}$ with $z = k$ and $\mathbb{S} = (1, 3, \ldots, 2k - 1, 2, 4, \ldots, 2k)$. Suppose that $n_p = m_p$ for all $p = 1, 2, \ldots, k$ (such tensors are often called even-order square tensors) and denote $\mathrm{GL}(\prod_{p=1}^{k} n_p, \mathbb{R})$ as the general linear group, the set of real-valued invertible matrices of size $\prod_{p=1}^{k} n_p \times \prod_{p=1}^{k} n_p$. Previous research [19, 37, 41, 55, 56] has proven that the mapping ψ is a group isomorphism from the even-order tensor space $\psi^{-1}(\mathrm{GL}(\prod_{p=1}^{k} n_p, \mathbb{R}))$ equipped with the Einstein product (1.9) to the general linear group. In other words, the space $\psi^{-1}(\mathrm{GL}(\prod_{p=1}^{k} n_p, \mathbb{R}))$ equipped with the Einstein product forms a group.

Using this unfolding ψ, numerous matrix notations and operations can be extended to even-order tensors. For example, the U-transpose (defined previously) can be defined alternatively as $\psi^{-1}(\psi(T)^{\top})$ for an even-order tensor T. More notations and operations, such as U-eigenvalues and tensor singular value decomposition, are developed based on the unfolding ψ.

1.4   Block Tensors

Block tensors can be defined similarly to block matrices [19, 40, 41]. Given two even-order tensors of the same size $T, S \in \mathbb{R}^{n_1 \times m_1 \times \cdots \times n_k \times m_k}$, the p-mode row block tensor, denoted by $\left[T \; S\right]_p \in \mathbb{R}^{n_1 \times m_1 \times \cdots \times n_p \times 2m_p \times \cdots \times n_k \times m_k}$, is defined as

$$\left(\left[T \; S\right]_p\right)_{j_1 i_1 \cdots j_p i_p \cdots j_k i_k} = \begin{cases} T_{j_1 i_1 \cdots j_p i_p \cdots j_k i_k} & \text{for } i_p = 1, 2, \ldots, m_p \\ S_{j_1 i_1 \cdots j_p i_p \cdots j_k i_k} & \text{for } i_p = m_p + 1, m_p + 2, \ldots, 2m_p \end{cases}. \tag{1.15}$$

The p-mode column block tensor, denoted by

$$\begin{bmatrix} T \\ S \end{bmatrix}_p \in \mathbb{R}^{n_1 \times m_1 \times \cdots \times 2n_p \times m_p \times \cdots \times n_k \times m_k},$$

can be defined similarly. The p-mode row/column block tensors possess many interesting properties that are similar to block matrices, e.g.,

- $P * [T\ S]_p = [P * T\ \ P * S]_p$ for $P \in \mathbb{R}^{s_1 \times n_1 \times \cdots \times s_k \times n_k}$;

- $\begin{bmatrix} T \\ S \end{bmatrix}_p * Q = \begin{bmatrix} T * Q \\ S * Q \end{bmatrix}_p$ for $Q \in \mathbb{R}^{m_1 \times r_1 \times \cdots \times m_k \times r_k}$;

- $[T\ S]_p * \begin{bmatrix} P \\ Q \end{bmatrix}_p = T * P + S * Q$ for $P, Q \in \mathbb{R}^{m_1 \times r_1 \times \cdots \times m_k \times r_k}$.

More importantly, the blocks of p-mode row/column block tensors can map to contiguous blocks under the unfolding ψ up to some permutations, i.e.,

$$\psi\left([T\ S]_p\right) = [\psi(T)\ \ \psi(S)]\mathbf{P} \quad \text{and} \quad \psi\left(\begin{bmatrix} T \\ S \end{bmatrix}_p\right) = \mathbf{Q}\begin{bmatrix} \psi(T) \\ \psi(S) \end{bmatrix}, \tag{1.16}$$

where $\mathbf{P}$ and $\mathbf{Q}$ are column and row permutation matrices [19, 40]. If $m_p = 1$ (or $n_p = 1$), $\mathbf{P}$ (or $\mathbf{Q}$) is the identity matrix.

Given s even-order tensors $T_p \in \mathbb{R}^{n_1 \times m_1 \times \cdots \times n_k \times m_k}$, one can obtain a p-mode row block tensor using (1.15) recursively. However, a more general concatenation approach can be defined for multiple blocks. Suppose that $s = s_1 s_2 \cdots s_k$. The generalized row block tensor, denoted by $[T_1\ T_2\ \cdots\ T_s] \in \mathbb{R}^{n_1 \times s_1 m_1 \times \cdots \times n_k \times s_k m_k}$, can be constructed in the following way:

- Compute the 1-mode row block tensors over $T_1, T_2, \ldots, T_s$ for every s_1 tensors and denote them by $T_1^{(1)}, T_2^{(1)}, \ldots, T_{s_2 s_3 \cdots s_k}^{(1)}$;
- Compute the 2-mode row block tensor over $T_1^{(1)}, T_2^{(1)}, \ldots, T_{s_2 s_3 \cdots s_k}^{(1)}$ for every s_2 tensors and denote them by $T_1^{(2)}, T_2^{(2)}, \ldots, T_{s_3 \cdots s_k}^{(2)}$;
- Keep repeating the process until the last k-mode row block tensor is obtained.

See Fig. 1.2 for an example. Generalized column block tensors can be constructed similarly. Note that the notation "$[\cdots]$" is used to represent the operations of both block matrices and generalized block tensors. Furthermore, the choice of factorization of s can affect the structure of generalized block tensors, which can be significant in tensor ranks/decompositions.

1.5 Tensor Eigenvalues

The study of eigenvalue problems for real supersymmetric tensors was initiated by Qi et al. [42, 57] and Lim [43] independently in 2005. Since then, various notions of tensor eigenvalues have been developed, which can be treated as generalizations of the matrix eigenvalue problem from different perspectives.

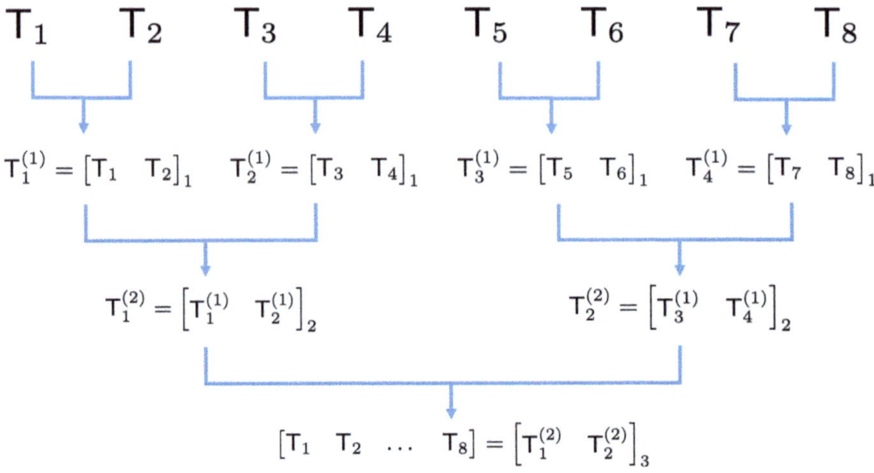

Fig. 1.2 An example of generalized row block tensor. This figure was redrawn from [19] with permission

1.5.1 H-Eigenvalues

Given a kth-order supersymmetric tensor $\mathsf{T} \in \mathbb{R}^{n \times n \times \overset{k}{\cdots} \times n}$, the eigenvalue $\lambda \in \mathbb{C}$ and eigenvector $\mathbf{v} \in \mathbb{C}^n$ of T are defined as

$$\mathsf{T}\mathbf{v}^{k-1} = \lambda \mathbf{v}^{[k-1]}, \tag{1.17}$$

where

$$\mathsf{T}\mathbf{v}^{k-1} = \mathsf{T} \times_1 \mathbf{v} \times_2 \mathbf{v} \times_3 \cdots \times_{k-1} \mathbf{v} \in \mathbb{R}^n,$$

and $\mathbf{v}^{[k-1]}$ denotes the elementwise $(k-1)$th power of $\mathbf{v}$. If λ is real, it is referred to as the H-eigenvalue of T (the corresponding eigenvector is called the H-eigenvector). Many classical matrix eigenvalue properties hold for this type of tensor eigenvalues. For example, the product of all the eigenvalues is equal to the determinant of T, the resultant of $\mathsf{T}\mathbf{v}^{k-1} = \mathbf{0}$. Similar to matrix eigenvalues, the maximum H-eigenvalue of T can be solved from the following optimization problem:

$$\max_{\mathbf{v} \in \mathbb{R}^n} \left\{ \mathsf{T}\mathbf{v}^k \;\middle|\; \sum_{j=1}^{n} v_j^k = 1 \right\}, \tag{1.18}$$

where v_j are the jth entries of $\mathbf{v}$. Since the objective function is continuous and the feasible set is compact, the solution of the optimization problem always exists [42]. The smallest H-eigenvalue of T can be found similarly.

1.5.2 Z-Eigenvalues

Given a kth-order supersymmetric tensor $\mathsf{T} \in \mathbb{R}^{n \times n \times \overset{k}{\cdots} \times n}$, the E-eigenvalue $\lambda \in \mathbb{C}$ and E-eigenvector $\mathbf{v} \in \mathbb{C}^n$ of T are defined as

$$\begin{cases} \mathsf{T}\mathbf{v}^{k-1} = \lambda \mathbf{v} \\ \mathbf{v}^{\top}\mathbf{v} = 1 \end{cases}. \tag{1.19}$$

When λ is real, it is referred to as the Z-eigenvalue of T (the corresponding eigenvector is called the Z-eigenvector). Similarly, the maximum Z-eigenvalue can be obtained by solving the following optimization problem:

$$\max_{\mathbf{v} \in \mathbb{R}^n} \left\{ \mathsf{T}\mathbf{v}^k \ \Big| \ \sum_{j=1}^{n} v_j^2 = 1 \right\}. \tag{1.20}$$

The smallest Z-eigenvalue of T can be found similarly. Additionally, computing the Z-eigenvalues (and H-eigenvalues) of a supersymmetric tensor is known to be NP-hard [58]. However, various methods have been proposed to compute or approximate them [59–61].

1.5.3 U-Eigenvalues

U-eigenvalues are defined based on the unfolding ψ for even-order square tensors [19, 37]. Given an even-order square tensor $\mathsf{T} \in \mathbb{R}^{n_1 \times n_1 \times \cdots \times n_k \times n_k}$, the U-eigenvalue $\lambda \in \mathbb{C}$ and U-eigentensor $\mathsf{V} \in \mathbb{C}^{n_1 \times n_2 \times \cdots \times n_k}$ are defined as

$$\mathsf{T} * \mathsf{V} = \lambda \mathsf{V}. \tag{1.21}$$

The notion of U-eigenvalues is a relaxation of Z-eigenvalues for even-order supersymmetric tensors. The optimization problem (1.20) can be rewritten as

$$\max \{ \mathsf{V}^{\top} * \mathsf{T} * \mathsf{V} \mid \|\mathsf{V}\| = 1 \text{ and } \mathsf{V} = \mathbf{v} \circ \mathbf{v} \circ \overset{k}{\cdots} \circ \mathbf{v} \}, \tag{1.22}$$

Therefore, the maximum Z-eigenvalue is always smaller than or equal to the maximum U-eigenvalue for an even-order supersymmetric tensor [21, 22]. Furthermore, unlike H-eigenvalues and Z-eigenvalues, U-eigenvalues can be computed efficiently through the eigenvalue decomposition of $\psi(\mathsf{T})$.

1.6 Tensor Decompositions

Tensor decompositions were first introduced in 1927 by Hitchcock [62, 63], but it was not until the 1960s that Tucker's works brought attention to this concept [64–66]. Since then, various types of tensor decompositions have been proposed, and they have become essential tools in many real-world applications [39].

1.6.1 Higher-Order Singular Value Decomposition

Higher-order singular value decomposition (HOSVD) is a multilinear analogue of matrix singular value decomposition (SVD) [67, 68]. Given a kth-order tensor $\mathsf{T} \in \mathbb{R}^{n_1 \times n_2 \times \cdots \times n_k}$, it can be decomposed as

$$\mathsf{T} = \mathsf{S} \times \{\mathbf{U}_1, \mathbf{U}_2, \ldots, \mathbf{U}_k\}, \tag{1.23}$$

where $\mathbf{U}_p \in \mathbb{R}^{n_p \times n_p}$ are orthogonal matrices and $\mathsf{S} \in \mathbb{R}^{n_1 \times n_2 \times \cdots \times n_k}$ is called the core tensor, see Fig. 1.3. The HOSVD (1.23) has a matricization form, i.e., the p-mode matricization of T can be represented as

$$\mathbf{T}_{(p)} = \mathbf{U}_p \mathbf{S}_{(p)} (\mathbf{U}_k \otimes \cdots \otimes \mathbf{U}_{p+1} \otimes \mathbf{U}_{p-1} \otimes \cdots \otimes \mathbf{U}_1)^\top,$$

where $\mathbf{T}_{(p)}$ and $\mathbf{S}_{(p)}$ are the p-mode unfoldings of T and S, respectively.

More interestingly, the sub-tensors $\mathsf{S}_{j_p=\alpha}$ of S (obtained by fixing the pth mode index to α) satisfy the following properties:

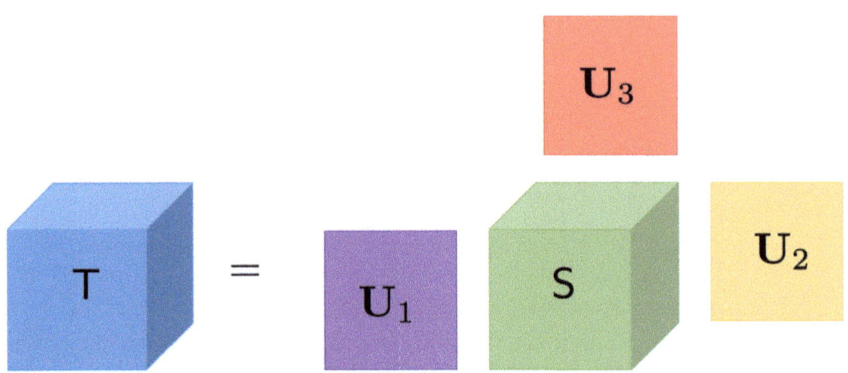

Fig. 1.3 An example of the HOSVD of a third-order tensor

- Two sub-tensors $\mathsf{S}_{j_p=\alpha}$ and $\mathsf{S}_{j_p=\beta}$ are orthogonal (i.e., $\langle \mathsf{S}_{j_p=\alpha}, \mathsf{S}_{j_p=\beta} \rangle = 0$) for $p = 1, 2, \ldots, k$ with $\alpha \neq \beta$;
- $\|\mathsf{S}_{j_p=1}\| \geq \|\mathsf{S}_{j_p=2}\| \geq \cdots \geq \|\mathsf{S}_{j_p=n_p}\|$ for $p = 1, 2, \ldots, k$.

The norms $\|\mathsf{S}_{j_p=\alpha}\|$ are referred to as the p-mode singular values of T. The number of non-vanishing p-mode singular values is called the p-rank of T. Previous research [67] has shown that the p-mode singular values are equal to the singular values of the p-mode unfolding. Therefore, the p-rank of T is equal to the rank of the p-mode unfolding.

Unlike matrix SVD, truncating the p-mode singular values only provides a suboptimal approximation of the original tensor [67]. To obtain the orthogonal matrices $\mathbf{U}_p$ (and the resulting HOSVD), one can simply compute the left singular vectors of the p-mode unfoldings. The core tensor S therefore can be recovered from the following:

$$\mathsf{S} = \mathsf{T} \times_1 \mathbf{U}_1^\top \times_2 \mathbf{U}_2^\top \times_3 \cdots \times_k \mathbf{U}_k^\top.$$

1.6.2 Tensor Singular Value Decomposition

Tensor singular value decomposition (TSVD) is defined based on the unfolding ψ for even-order tensors [55, 69–71]. For an even-order tensor $\mathsf{T} \in \mathbb{R}^{n_1 \times m_1 \times \cdots \times n_k \times m_k}$, the TSVD of T is defined as

$$\mathsf{T} = \mathsf{U} * \mathsf{S} * \mathsf{V}^\top, \tag{1.24}$$

where $\mathsf{U} \in \mathbb{R}^{n_1 \times n_1 \times \cdots \times n_k \times n_k}$ and $\mathsf{V} \in \mathbb{R}^{m_1 \times m_1 \times \cdots \times m_k \times m_k}$ are U-orthogonal tensors (i.e., $\psi(\mathsf{U})$ and $\psi(\mathsf{V})$ are orthogonal matrices), and $\mathsf{S} \in \mathbb{R}^{n_1 \times m_1 \times \cdots \times n_k \times m_k}$ is a U-diagonal tensor (i.e., $\psi(\mathsf{S})$ is a diagonal matrix) including the U-singular values of T along the diagonal. Obviously, the TSVD (1.24) has a matricization form, i.e.,

$$\psi(\mathsf{T}) = \psi(\mathsf{U})\psi(\mathsf{S})\psi(\mathsf{V})^\top.$$

Therefore, the TSVD can be solved readily from the SVD of $\psi(\mathsf{T})$.

Suppose that the rank of $\psi(\mathsf{T})$ is equal to r (also known as the unfolding rank of T). The TSVD of T can be written in the following economy-size form as:

$$\mathsf{T} = \sum_{j=1}^{r} \sigma_j \mathsf{X}_j * \mathsf{Y}_j^\top, \tag{1.25}$$

where $\mathsf{X}_j \in \mathbb{R}^{n_1 \times 1 \times \cdots \times n_k \times 1}$ and $\mathsf{Y}_j \in \mathbb{R}^{m_1 \times 1 \times \cdots \times m_k \times 1}$ can be viewed as the jth columns of $\psi(\mathsf{U})$ and $\psi(\mathsf{V})$ under ψ^{-1}, respectively. The U-singular values are arranged in descending order. Suppose that $r = r_1 r_2 \cdots r_k$. The economy-size TSVD (1.25) can be rewritten in the form of (1.24) by using generalized row block tensors, i.e., $\mathsf{U} \in \mathbb{R}^{n_1 \times r_1 \times \cdots \times n_k \times r_k}$,

$S \in \mathbb{R}^{r_1 \times r_1 \times \cdots \times r_k \times r_k}$, and $V \in \mathbb{R}^{m_1 \times r_1 \times \cdots \times m_k \times r_k}$. It is worth noting that TSVD is a special case of HOSVD according to (1.10) for even-order tensors.

1.6.3 CANDECOMP/PARAFAC Decomposition

CANDECOMP/PARAFAC decomposition (CPD) can be seen as a higher-order generalization of matrix eigenvalue decomposition [39]. Given a kth-order tensor $T \in \mathbb{R}^{n_1 \times n_2 \times \cdots \times n_k}$, it can be decomposed into the sum of rank-one tensors as follows:

$$T = \sum_{j=1}^{r} \lambda_j \mathbf{V}_{:j}^{(1)} \circ \mathbf{V}_{:j}^{(2)} \circ \cdots \circ \mathbf{V}_{:j}^{(k)}, \tag{1.26}$$

where $\lambda_j \in \mathbb{R}^+$ are weights, $\mathbf{V}^{(p)} \in \mathbb{R}^{n_p \times r}$ are factor matrices with unit length columns, and r is called the CP rank if it is the minimum integer that achieves the decomposition, see Fig. 1.4A. The CPD (1.26) also has a matricization form, i.e., the p-mode matricization of T can be represented as

$$\mathbf{T}_{(p)} = \mathbf{V}^{(p)}\mathbf{\Lambda}(\mathbf{V}^{(k)} \odot \cdots \odot \mathbf{V}^{(p+1)} \odot \mathbf{V}^{(p-1)} \odot \cdots \odot \mathbf{V}^{(1)}),$$

where $\mathbf{T}_{(p)}$ is the p-mode unfolding of T and $\mathbf{\Lambda}$ is a diagonal matrix that includes λ_j along the diagonal.

The CPD (1.26) can be solved using alternating least squares methods, and it is unique up to permutation and scaling under a weak condition, i.e.,

$$\sum_{p=1}^{k} \mathrm{Krank}_{\mathbf{V}^{(p)}} \geq 2r + k - 1,$$

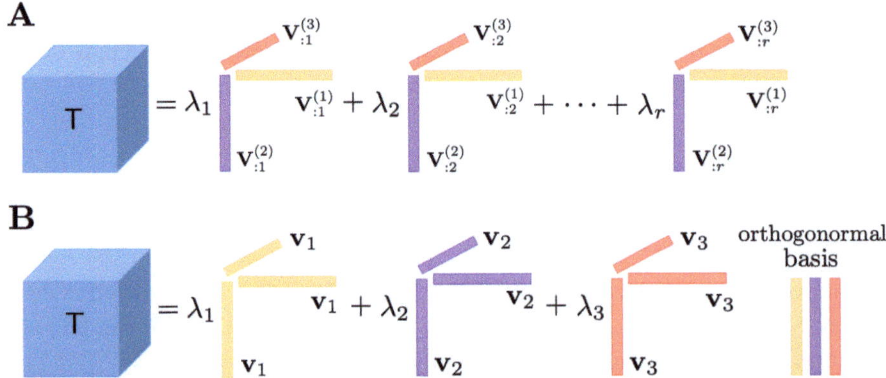

Fig. 1.4 Examples of the CPD (**a**) and the orthogonal decomposition (**b**) of a third-order tensor

where Krank is the maximum number of columns of a matrix that are linearly independent of each other [39]. Although the best CP rank approximation is ill-posed, truncating the CP rank can still yield a good approximation of the original tensor [19, 39]. It is worth noting that both CPD and HOSVD can be expressed as Tucker products, known as Tucker decomposition (TD).

CPD can be extended to even-order tensors in a generalized form known as generalized CPD [19, 72]. Given an even-order tensor $\mathsf{T} \in \mathbb{R}^{n_1 \times m_1 \times \cdots \times n_k \times m_k}$, it can be decomposed as

$$\mathsf{T} = \sum_{j=1}^{r} \lambda_j \mathsf{V}^{(1)}_{::j} \circ \mathsf{V}^{(2)}_{::j} \circ \cdots \circ \mathsf{V}^{(k)}_{::j}, \tag{1.27}$$

where $\lambda_j \in \mathbb{R}^+$ are weights, $\mathsf{V}^{(p)} \in \mathbb{R}^{n_p \times m_p \times r}$ are third-order factor tensors, and r is called the generalized CP rank (also called Kronecker rank) if it is the minimum integer that achieves the decomposition.

If two even-order tensors are given in the generalized CPD form, the Einstein product between the two can be computed efficiently without using (1.8). Given two even-order tensors $\mathsf{T} \in \mathbb{R}^{n_1 \times m_1 \times \cdots \times n_k \times m_k}$ and $\mathsf{S} \in \mathbb{R}^{m_1 \times h_1 \times \cdots \times m_k \times h_k}$, both in the generalized CPD form with factor tensors $\mathsf{V}^{(p)}$ and $\mathsf{U}^{(p)}$, and with generalized CP ranks r and s, respectively, the Einstein product can be computed in the generalized CPD form as follows:

$$\mathsf{T} * \mathsf{S} = \sum_{l=1}^{t} \mathsf{W}^{(1)}_{::l} \circ \mathsf{W}^{(2)}_{::l} \circ \cdots \circ \mathsf{W}^{(k)}_{::l}, \tag{1.28}$$

where $\mathsf{W}^{(p)}_{::l} = \mathsf{V}^{(p)}_{::j} \mathsf{U}^{(p)}_{::i} \in \mathbb{R}^{n_p \times h_p}$ for $l = \varphi(\{j, i\}, \{r, s\})$ and $t = rs$.

Remark 1.1 Assuming $n_p = m_p = h_p = n$ and $r = s$, the time complexity of computing the Einstein product in the generalized CPD form is approximately $\mathcal{O}(kn^3 r)$. In contrast, using (1.8) to compute the Einstein product requires around $\mathcal{O}(n^{3k})$ operations.

1.6.4 Tensor Orthogonal Decomposition

Tensor orthogonal decomposition is a special case of CPD that only applies to supersymmetric tensors [73]. It's worth noting that not all supersymmetric tensors have the orthogonal decomposition. Given a kth-order supersymmetric tensor $\mathsf{T} \in \mathbb{R}^{\overset{k}{\overbrace{n \times n \times \cdots \times n}}}$, the orthogonal decomposition decomposes T as

$$\mathsf{T} = \sum_{j=1}^{n} \lambda_j \mathbf{v}_j \circ \mathbf{v}_j \circ \overset{k}{\cdots} \circ \mathbf{v}_j, \tag{1.29}$$

where $\mathbf{v}_j \in \mathbb{R}^n$ are orthonormal vectors, as shown in Fig. 1.4b. Previous research [73] has shown that λ_j are the Z-eigenvalues of T with the corresponding Z-eigenvectors $\mathbf{v}_j$ (though λ_j do not include all the Z-eigenvalues of T). If a supersymmetric tensor can be decomposed as in (1.29), it is called orthogonally decomposable (odeco). Odeco tensors possess the desirable orthonormal property, which can be applied to various applications, such as estimating parameters in the method of moments from statistics [74]. While tensor orthogonal decomposition is NP-hard like CPD, a tensor-based power method [73] has been developed to obtain the orthogonal decomposition of an odeco tensor. It is important to note that tensor orthogonal decomposition is also a specific case of HOSVD, where S is a diagonal tensor with λ_j as its diagonal entries.

1.6.5 Tensor Train Decomposition

Tensor train decomposition (TTD) is similar to CPD, but can offer better numerical stability. The TTD decomposes a kth-order tensor $T \in \mathbb{R}^{n_1 \times n_2 \times \cdots \times n_k}$ in the following way:

$$
T = \sum_{j_0=1}^{r_0} \sum_{j_1=1}^{r_1} \cdots \sum_{j_k=1}^{r_k} V_{j_0:j_1}^{(1)} \circ V_{j_1:j_2}^{(2)} \circ \cdots \circ V_{j_{k-1}:j_k}^{(k)}, \tag{1.30}
$$

where $\{r_0, r_1, \ldots, r_k\}$ is the set of TT-ranks with $r_0 = r_k = 1$, and $V^{(p)} \in \mathbb{R}^{r_{p-1} \times n_p \times r_p}$ are third-order factor tensors, see Fig. 1.5. Since the TTD is numerically stable, it can return optimal TT-ranks, which are computed as

$$
r_p = \text{rank}(\mathbf{T}_{(p,\{1,2,\ldots,k\})}),
$$

where $\mathbf{T}_{(p,\{1,2,\ldots,k\})}$ are the unfolded matrices of T. A factor tensor $V^{(p)}$ is called left-orthonormal if it satisfies the following condition:

$$
(\mathbf{V}_{(2,\{1,2,3\})}^{(p)})^{\top} \mathbf{V}_{(2,\{1,2,3\})}^{(p)} = \mathbf{I} \in \mathbb{R}^{n_p \times n_p},
$$

and is called right-orthonormal if it satisfies the following condition:

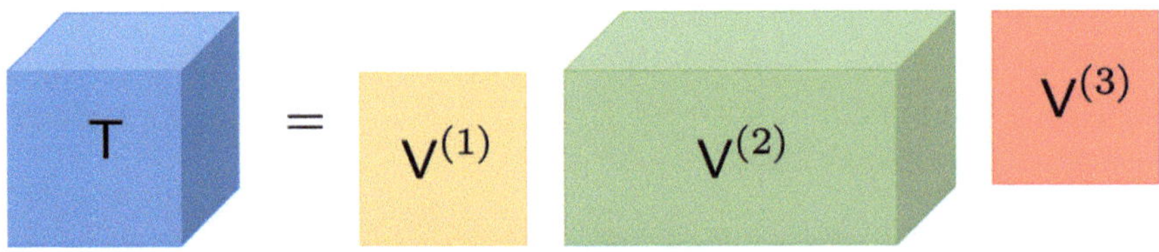

Fig. 1.5 An example of the TTD of a third-order tensor

$$\mathbf{V}^{(p)}_{(1,\{1,2,3\})}(\mathbf{V}^{(p)}_{(1,\{1,2,3\})})^\top = \mathbf{I} \in \mathbb{R}^{n_{p-1}\times n_{p-1}},$$

where $\mathbf{V}^{(p)}_{(2,\{1,2,3\})}$ and $\mathbf{V}^{(p)}_{(1,\{1,2,3\})}$ are the left-/right-unfolded matrices of $\mathbf{V}^{(p)}$. Here, $\mathbf{I}$ denotes the identity matrices.

TTD has several advantages over CPD. First, it can provide better approximation, i.e., truncating the TT-ranks is quasi-optimal. Additionally, TTD is computationally robust. One interesting application of TTD is solving the TSVD of an even-order tensor efficiently, without the need for the unfolding ψ [69, 75], see Algorithm 1. Note that in Step 2, if the TTD of T is provided, the TTD of $\bar{\mathsf{T}}$ can be computed without converting to the full representation. In Step 5, the `reshape` operation refers to the MATLAB operation used for reshaping tensors.

Remark 1.2 Assuming that $n_p = m_p = n$ and the optimal TT-ranks of $\bar{\mathsf{T}}$ are equal to r, the economy-size TSVD via TTD has a time complexity of $\mathcal{O}(knr^3)$ and a space complexity of $\mathcal{O}(knr^2)$. In contrast, the matrix SVD-based TSVD through unfolding requires around

$$\mathcal{O}\left(\min\left\{\prod_{p=1}^{k} n_p^2 m_p,\ \prod_{p=1}^{k} n_p m_p^2\right\}\right)$$

number of operations.

Similarly, TTD can also be defined in a generalized form for even-order tensors (referred to as generalized TTD). Given an even-order tensor $\mathsf{T} \in \mathbb{R}^{n_1\times m_1\times\cdots\times n_k\times m_k}$, it can be decomposed as

$$\mathsf{T} = \sum_{j_0=1}^{r_0}\sum_{j_1=1}^{r_1}\cdots\sum_{j_k=1}^{r_k} \mathsf{V}^{(1)}_{j_0::j_1} \circ \mathsf{V}^{(2)}_{j_1::j_2} \circ\cdots\circ \mathsf{V}^{(k)}_{j_{k-1}::j_k}, \tag{1.31}$$

where $\{r_0, r_1, \ldots, r_k\}$ is the set of the generalized TT-ranks with $r_0 = r_k = 1$, and $\mathsf{V}^{(p)} \in \mathbb{R}^{r_{p-1}\times n_p\times m_p\times r_p}$ are fourth-order factor tensors. To further reduce the complexity of generalized TTD, one can consider quantizing each dimension of the tensor by setting

$$n_p = \prod_{q=1}^{b_p} n_{pq} \quad \text{and} \quad m_p = \prod_{q=1}^{b_p} m_{pq}$$

for some positive integers b_p [25, 72, 76]. This extension is referred to as quantized TTD, and a typical choice for n_{pq} and m_{pq} is 2. It's worth noting that the quantized TTD of T is equivalent to the generalized TTD of a reshaped tensor of T, denoted by

$$\tilde{\mathsf{T}} \in \mathbb{R}^{n_{11}\times m_{11}\times\cdots\times n_{1b_1}\times m_{1b_1}\times\cdots\times n_{k1}\times m_{k1}\times\cdots\times n_{kb_k}\times m_{kb_k}}.$$

Algorithm 1 Economy-size TSVD via TTD. This algorithm was adapted from [69] with permission.

1: Given an even-order tensor $\mathsf{T} \in \mathbb{R}^{n_1 \times m_1 \times \cdots \times n_k \times m_k}$

2: Find the $\mathbb{S}$-transpose of T with $\mathbb{S} = \{1, 3, \ldots, 2k-1, 2, 4, \ldots, 2k\}$, denoted by $\bar{\mathsf{T}} \in \mathbb{R}^{n_1 \times n_2 \times \cdots \times n_k \times m_1 \times m_2 \times \cdots \times m_k}$ and compute the TTD of $\bar{\mathsf{T}}$

3: Left-orthonormalize the first k core tensors of $\bar{\mathsf{T}}$ and right-orthonormalize the last $k-1$ core tensors of $\bar{\mathsf{T}}$

4: Compute the economy-size matrix SVD of the kth left-unfolded matrix of $\bar{\mathsf{T}}$, i.e.,

$$\mathsf{V}^{(k)}_{(2,\{1,2,3\})} = \mathbf{U}\boldsymbol{\Sigma}\mathbf{V}^\top = \sum_{j=1}^{r} \sigma_j \mathbf{u}_j \mathbf{v}_j^\top,$$

where r is the rank of the matrix.

5: Set $\mathsf{V}^{(k)} = \texttt{reshape}(\mathbf{U}, r_{k-1}, n_k, r)$ and $\mathsf{V}^{(k+1)} = \texttt{reshape}(\mathbf{V}^\top \mathsf{V}^{(k+1)}_{(1,\{1,2,3\})}, r, m_1, r_{k+1})$

6: The left- and right-singular tensors of T are given by

$$\mathsf{X}_j = \sum_{j_0=1}^{r_0} \sum_{j_1=1}^{r_1} \cdots \sum_{j_{k-1}=1}^{r_{k-1}} \mathsf{V}^{(1)}_{j_0:j_1} \circ \mathsf{V}^{(2)}_{j_1:j_2} \circ \cdots \circ \mathsf{V}^{(k)}_{j_{k-1}:j},$$

$$\mathsf{Y}_j = \sum_{j_k=1}^{r_k} \sum_{j_{k+1}=1}^{r_{k+1}} \cdots \sum_{j_{2k}=1}^{r_{2k}} \mathsf{X}^{(k+1)}_{j:j_{k+1}} \circ \mathsf{X}^{(k+2)}_{j_{k+1}:j_{k+2}} \circ \cdots \circ \mathsf{X}^{(2k)}_{j_{2k-1}:j_{2k}},$$

respectively

7: **return** Left- and right-singular tensors X_j and Y_j with U-singular values σ_j of T.

The Einstein product can be computed efficiently in the generalized TTD form analogous to (1.28). Given two even-order tensors $\mathsf{T} \in \mathbb{R}^{n_1 \times m_1 \times \cdots \times n_k \times m_k}$ and $\mathsf{S} \in \mathbb{R}^{m_1 \times h_1 \times \cdots \times m_k \times h_k}$ in the generalized TTD form with factor tensors $\mathsf{V}^{(p)}$ and $\mathsf{U}^{(p)}$, and with generalized TTD ranks $\{r_0, r_1, \ldots, r_k\}$ and $\{s_0, s_1, \ldots, s_k\}$, respectively, the Einstein product is then computed as

$$\mathsf{T} * \mathsf{S} = \sum_{l_0=1}^{t_0} \sum_{l_1=1}^{t_1} \cdots \sum_{l_k=1}^{t_k} \mathsf{W}^{(1)}_{l_0::l_1} \circ \mathsf{W}^{(2)}_{l_1::l_2} \circ \cdots \circ \mathsf{W}^{(k)}_{l_{k-1}::l_k}, \tag{1.32}$$

where

$$\mathsf{W}^{(p)}_{l_{p-1}::l_p} = \mathsf{V}^{(p)}_{j_{p-1}::j_p} \mathsf{U}^{(p)}_{i_{p-1}::i_p} \in \mathbb{R}^{n_p \times h_p}$$

for $l_p = \varphi(\{j_p, i_p\}, \{r_p, s_p\})$ and $t_p = r_p s_p$.

Remark 1.3 The computational complexity of the Einstein product in the generalized TTD form is about $\mathcal{O}(kn^3 r^4)$ (assuming $n_p = m_p = h_p = n$ and $r_p = s_p = r$).

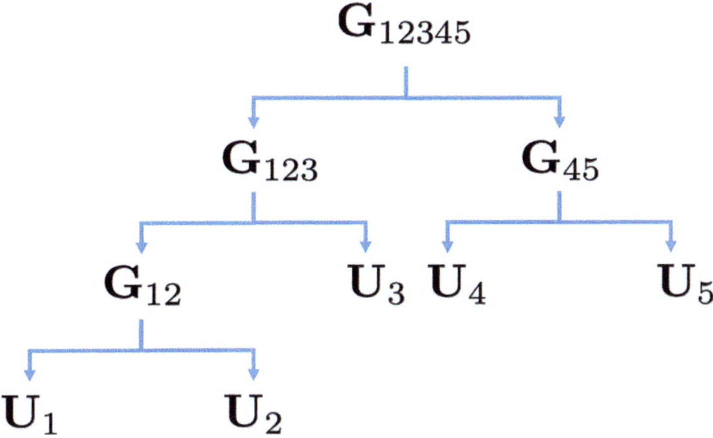

Fig. 1.6 An example of a binary tree of a fifth-order tensor

In addition, a range of other tensor operations, including summation, outer/inner product, U-eigenvalue, U-transpose, and solving multilinear equations (i.e., $\mathsf{T} * \mathsf{X} = \mathsf{B}$), can also be efficiently performed using the generalized TTD form. For further details, interested readers may refer to [69, 72]. It's worth noting that all these results also apply to quantized TTD.

1.6.6 Hierarchical Tucker Decomposition

Hierarchical Tucker decomposition (HTD) is a structured tensor representation method that efficiently approximates a high-dimensional tensor using a multi-level approach [77–79]. In fact, TTD is a specific instance of HTD. The general idea behind HTD is to recursively split the modes of a tensor $\mathsf{T} \in \mathbb{R}^{n_1 \times n_2 \times \cdots \times n_k}$, resulting in a binary tree structure for its dimensions (Fig. 1.6). The binary tree captures a hierarchy of matrices $\mathbf{U}_p$, which form a basis for the column spaces of the p-mode matricization $\mathbf{T}_{(p)}$ for $p \subset \{1, 2, \ldots, k\}$. Not all matrices in the HTD of T need to be explicitly stored, as the matrix $\mathbf{U}_p$ of a parent node can be computed from its left child $\mathbf{U}_{pl}$ and right child $\mathbf{U}_{pr}$ as

$$\mathbf{U}_p = (\mathbf{U}_{pl} \otimes \mathbf{U}_{pr})\mathbf{G}_p, \tag{1.33}$$

where $\mathbf{G}_p \in \mathbb{R}^{r_{pl}r_{pr} \times r_p}$ are called the transfer matrices with HT-ranks r_{pl}, r_{pr}, and r_p at nodes p_l, p_r, and p, respectively. The construction of the HTD begins with the leaf nodes and proceeds recursively by applying (1.33) until the root of the binary tree is reached. It is important to note that the rank r_p is fixed to 1 at the root node. HTD is particularly useful when dealing with very large tensors, as it allows for a compact representation that can be stored and manipulated efficiently.

1.6.7 t-Singular Value/Eigenvalue Decomposition

t-singular value decomposition (t-SVD) is a generalization of matrix SVD for third-order tensors [45]. The t-SVD decomposes a third-order tensor $\mathsf{T} \in \mathbb{R}^{n \times m \times s}$ as

$$\mathsf{T} = \mathsf{U} \star \mathsf{S} \star \mathsf{V}^\top, \tag{1.34}$$

where $\mathsf{U} \in \mathbb{R}^{n \times n \times s}$ and $\mathsf{V} \in \mathbb{R}^{m \times m \times s}$ are t-orthogonal, and $\mathsf{S} \in \mathbb{R}^{n \times m \times s}$ is f-diagonal (each frontal slice of S is a diagonal matrix). The tubes of S, i.e., $\mathsf{S}_{jj:}$, are referred to as the singular tuples of T. The t-SVD (1.34) can be rewritten in the economy-size form as:

$$\mathsf{T} = \sum_{j=1}^{r} \mathsf{U}_{:j:} \star \mathsf{S}_{jj:} \star \mathsf{V}_{:j:}^\top, \tag{1.35}$$

where r is called the t-rank of T defined as the number of nonzero singular tuples.

Similarly, the t-eigenvalue decomposition decomposes a third-order tensor $\mathsf{T} \in \mathbb{R}^{n \times n \times s}$ as

$$\mathsf{T} = \mathsf{V} \star \mathsf{D} \star \mathsf{V}^{-1}, \tag{1.36}$$

where $\mathsf{V} \in \mathbb{C}^{n \times n \times s}$ and $\mathsf{D} \in \mathbb{C}^{n \times n \times s}$ is f-diagonal. The tubes of D, i.e., $\mathsf{D}_{jj:}$, are referred to as the eigentuples.

The t-SVD (or t-eigenvalue decomposition) can be computed using the matrix SVD (or eigenvalue decomposition) in the Fourier domain, i.e.,

$$(\mathbf{F}_n \otimes \mathbf{I}_s)\texttt{bcirc}(\mathsf{T})(\mathbf{F}_n^\top \otimes \mathbf{I}_s) = \texttt{blockdiag}(\mathbf{T}_1, \mathbf{T}_2, \ldots, \mathbf{T}_s), \tag{1.37}$$

where $\texttt{blockdiag}$ denotes the MATLAB function that creates a block diagonal matrix, $\mathbf{F}_n \in \mathbb{R}^{n \times n}$ is the discrete Fourier transform matrix defined as

$$\mathbf{F}_n = \frac{1}{\sqrt{n}} \begin{bmatrix} 1 & 1 & 1 & \cdots & 1 \\ 1 & \omega & \omega^2 & \cdots & \omega^{n-1} \\ \vdots & \vdots & \vdots & \ddots & \vdots \\ 1 & \omega^{n-1} & \omega^{2(n-1)} & \cdots & \omega^{(n-1)^2} \end{bmatrix}$$

with $\omega = \exp\{\frac{-2\pi j}{n}\}$ (note that j denotes the imaginary number here), and $\mathbf{I}_s \in \mathbb{R}^{s \times s}$ denotes the identity matrix of size s. Then the t-SVD/eigenvalue decomposition of T can be constructed (implicitly) by applying the appropriate matrix decomposition to $\mathbf{T}_j$. More details can be found in [44, 45, 47].

References

1. Voigt, Woldemar. 1898. *Die fundamentalen physikalischen Eigenschaften der Krystalle in elementarer Darstellung*. de Gruyter.
2. Irgens, Fridtjov. 2008. *Continuum mechanics*. Springer Science & Business Media.
3. Spencer, Anthony James Merrill. 2004. *Continuum mechanics*. Courier Corporation.
4. Gurtin, Morton E. 1982. *An introduction to continuum mechanics*. Academic.
5. Katzin, Gerald H., Jack Levine, and William R. Davis. 1969. Curvature collineations: A fundamental symmetry property of the space-times of general relativity defined by the vanishing lie derivative of the riemann curvature tensor. *Journal of Mathematical Physics* 10 (4): 617–629.
6. Kühnel, Wolfgang. 2015. *Differential geometry*, vol. 77. American Mathematical Society.
7. Guggenheimer, Heinrich W. (2012). *Differential geometry*. Courier Corporation.
8. Einstein, Albert. 1922. The general theory of relativity. In *The meaning of relativity*, 54–75. Berlin: Springer.
9. Hawking, Stephen and W. Israel. 2010. General relativity: an einstein centenary survey. *General relativity: An Einstein Centenary Survey.*
10. Cammoun, Leila, Carlos-Alberto Castaño-Moraga, Emma Muñoz-Moreno, Darío Sosa-Cabrera, Burak Acar, Miguel A Rodriguez-Florido, Anders Brun, Hans Knutsson and Jean-Philippe Thiran. 2009. A review of tensors and tensor signal processing. In *Tensors in image processing and computer vision*, 1–32.
11. Sidiropoulos, Nicholas D, Lieven De Lathauwer, Xiao Fu, Kejun Huang, Evangelos E Papalexakis, and Christos Faloutsos. 2017. Tensor decomposition for signal processing and machine learning. *IEEE Transactions on Signal Processing* 65 (13): 3551–3582.
12. Favier, Gérard. 2021. *Matrix and Tensor Decompositions in Signal Processing*, vol. 2. New York: Wiley.
13. Ji, Yuwang, Qiang Wang, Xuan Li, and Jie Liu. 2019. A survey on tensor techniques and applications in machine learning. *IEEE Access* 7: 162950–162990.
14. Signoretto, Marco, Quoc Tran Dinh, Lieven De Lathauwer, and Johan AK Suykens. Learning with tensors: a framework based on convex optimization and spectral regularization. *Machine Learning*, 94 (3): 303–351.
15. Rabanser, Stephan, Oleksandr Shchur, and Stephan Günnemann. 2017. Introduction to tensor decompositions and their applications in machine learning. arXiv:1711.10781.
16. Bi, Xuan, and Xiwei Tang. 2021. Tensors in statistics. *Annual Review of Statistics and Its Application*, 8.
17. Koutsoukos, Dimitrios, Supun Nakandala, Konstantinos Karanasos, Karla Saur, Gustavo Alonso, and Matteo Interlandi. 2021. Tensors: An abstraction for general data processing. *Proceedings of the VLDB Endowment* 14 (10): 1797–1804.
18. Bocci, Cristiano, and Luca Chiantini. *An Introduction to Algebraic Statistics with Tensors*, vol. 1. Berlin: Springer.
19. Chen, Can, Amit Surana, Anthony M Bloch, and Indika Rajapakse. 2021. Multilinear control systems theory. *SIAM Journal on Control and Optimization* 59 (1): 749–776.
20. Rogers, Mark, Lei Li, and Stuart J Russell. 2013. Multilinear dynamical systems for tensor time series. *Advances in Neural Information Processing Systems*, 26.
21. Chen, Can. 2023. Explicit solutions and stability properties of homogeneous polynomial dynamical systems. *IEEE Transactions on Automatic Control* 68 (8): 4962–4969.
22. Chen, Can. 2024. On the stability of discrete-time homogeneous polynomial dynamical systems. *Computational and Applied Mathematics* 43: 75.
23. Lim, Lek-Heng. 2021. Tensors in computations. *Acta Numerica* 30: 555–764.

24. Ballani, Jonas, and Lars Grasedyck. 2013. A projection method to solve linear systems in tensor format. *Numerical linear algebra with applications* 20 (1): 27–43.
25. Oseledets, Ivan V. 2010. Approximation of $2^d \times 2^d$ matrices using tensor decomposition. *SIAM Journal on Matrix Analysis and Applications* 31 (4): 2130–2145.
26. Aja-Fernández, Santiago, Rodrigo de Luis Garcia, Dacheng Tao, and Xuelong Li. 2009. *Tensors in Image Processing and Computer Vision*. Springer Science & Business Media.
27. Panagakis, Yannis, Jean Kossaifi, Grigorios G Chrysos, James Oldfield, Mihalis A Nicolaou, Anima Anandkumar, and Stefanos Zafeiriou. 2021. Tensor methods in computer vision and deep learning. *Proceedings of the IEEE* 109 (5): 863–890.
28. Shashua, Amnon, and Tamir Hazan. 2005. Non-negative tensor factorization with applications to statistics and computer vision. In *Proceedings of the 22nd international conference on machine learning*, 792–799.
29. Mori, Susumu, and Jiangyang Zhang. 2006. Principles of diffusion tensor imaging and its applications to basic neuroscience research. *Neuron* 51 (5): 527–539.
30. Soares, José M, Paulo Marques, Victor Alves, and Nuno Sousa. 2013. A hitchhiker's guide to diffusion tensor imaging. *Frontiers in Neuroscience*, 7: 31.
31. Cong, Fengyu, Qiu-Hua. Lin, Li-Dan. Kuang, Xiao-Feng. Gong, Piia Astikainen, and Tapani Ristaniemi. 2015. Tensor decomposition of eeg signals: a brief review. *Journal of Neuroscience Methods* 248: 59–69.
32. Chen, Haibin, Yannan Chen, Guoyin Li, and Liqun Qi. 2018. A semidefinite program approach for computing the maximum eigenvalue of a class of structured tensors and its applications in hypergraphs and copositivity test. *Numerical Linear Algebra with Applications* 25 (1): e2125.
33. Chen, Can, and Indika Rajapakse. 2020. Tensor entropy for uniform hypergraphs. *IEEE Transactions on Network Science and Engineering* 7 (4): 2889–2900.
34. Xie, Jinshan, and An. Chang. 2013. On the z-eigenvalues of the adjacency tensors for uniform hypergraphs. *Linear Algebra and Its Applications* 439 (8): 2195–2204.
35. Chen, Can, Amit Surana, Anthony Bloch, and Indika Rajapakse. 2019. Multilinear time invariant system theory. In *2019 Proceedings of the conference on control and its applications*, 118–125. SIAM.
36. Chen, Haiming, Jie Chen, Lindsey A Muir, Scott Ronquist, Walter Meixner, Mats Ljungman, Thomas Ried, Stephen Smale, and Indika Rajapakse. 2015. Functional organization of the human 4d nucleome. *Proceedings of the National Academy of Sciences* 112 (26): 8002–8007.
37. Cui, Lu-Bin, Chuan Chen, Wen Li, and Michael K Ng. 2016. An eigenvalue problem for even order tensors with its applications. *Linear and Multilinear Algebra* 64 (4): 602–621.
38. Sang, Caili, and Zhen Chen. 2020. Z-eigenvalue localization sets for even order tensors and their applications. *Acta Applicandae Mathematicae* 169 (1): 323–339.
39. Kolda, Tamara G, and Brett W Bader. 2009. Tensor decompositions and applications. *SIAM Review* 51 (3): 455–500.
40. Ragnarsson, Stefan, and Charles F Van Loan. 2012. Block tensor unfoldings. *SIAM Journal on Matrix Analysis and Applications* 33 (1): 149–169.
41. Brazell, Michael, Na. Li, Carmeliza Navasca, and Christino Tamon. 2013. Solving multilinear systems via tensor inversion. *SIAM Journal on Matrix Analysis and Applications* 34 (2): 542–570.
42. Qi, Liqun. 2005. Eigenvalues of a real supersymmetric tensor. *Journal of Symbolic Computation* 40 (6): 1302–1324.
43. Lim, Lek-Heng. 2005. Singular values and eigenvalues of tensors: a variational approach. In *1st IEEE international workshop on computational advances in multi-sensor adaptive processing*, 129–132. IEEE.
44. Kilmer, Misha E, Karen Braman, Ning Hao, and Randy C Hoover. 2013. Third-order tensors as operators on matrices: A theoretical and computational framework with applications in imaging. *SIAM Journal on Matrix Analysis and Applications* 34 (1): 148–172.

45. Kilmer, Misha E, Carla D Martin, and Lisa Perrone. 2008. A third-order generalization of the matrix svd as a product of third-order tensors. *Tufts University, Department of Computer Science, Technical Report. TR-2008-4.*

46. Kilmer, Misha E, and Carla D Martin. 2011. Factorization strategies for third-order tensors. *Linear Algebra and Its Applications* 435 (3): 641–658.

47. Zhang, Zemin, and Shuchin Aeron. 2016. Exact tensor completion using t-svd. *IEEE Transactions on Signal Processing* 65 (6): 1511–1526.

48. Pollock, D. Stephen G. 2013. On kronecker products, tensor products and matrix differential calculus. *International Journal of Computer Mathematics* 90 (11): 2462–2476.

49. Liu, Shuangzhe, Gotz Trenkler, et al. 2008. Hadamard, khatri-rao, kronecker and other matrix products. *International Journal of Information and Systems Sciences* 4 (1): 160–177.

50. Broxson, Bobbi Jo. 2006. *The kronecker product.*

51. Pickard, Joshua, Cooper Stansbury, Can Chen, Amit Surana, Anthony Bloch, and Indika Rajapakse. 2023. Kronecker product of tensors and hypergraphs. arXiv:2305.03875.

52. Liu, Shuangzhe. 2002. Several inequalities involving khatri-rao products of positive semidefinite matrices. *Linear Algebra and Its Applications* 354 (1–3): 175–186.

53. Qi, Liqun. 2006. Rank and eigenvalues of a supersymmetric tensor, the multivariate homogeneous polynomial and the algebraic hypersurface it defines. *Journal of Symbolic Computation* 41 (12): 1309–1327.

54. Brachat, Jerome, Pierre Comon, Bernard Mourrain, and Elias Tsigaridas. 2010. Symmetric tensor decomposition. *Linear Algebra and Its Applications* 433 (11–12): 1851–1872.

55. Sun, Lizhu, Baodong Zheng, Bu. Changjiang, and Yimin Wei. 2016. Moore-penrose inverse of tensors via einstein product. *Linear and Multilinear Algebra* 64 (4): 686–698.

56. Liang, Mao-lin, Bing Zheng, and Rui-juan Zhao. 2019. Tensor inversion and its application to the tensor equations with einstein product. *Linear and Multilinear Algebra* 67 (4): 843–870.

57. Qi, Liqun. 2007. Eigenvalues and invariants of tensors. *Journal of Mathematical Analysis and Applications* 325 (2): 1363–1377.

58. Hillar, Christopher J., and Lek-Heng Lim. 2013, Most tensor problems are np-hard. *Journal of the ACM (JACM)* 60 (6): 1–39.

59. Ng, Michael, Liqun Qi, and Guanglu Zhou. 2010. Finding the largest eigenvalue of a nonnegative tensor. *SIAM Journal on Matrix Analysis and Applications* 31 (3): 1090–1099.

60. Chen, Liping, Lixing Han, and Liangmin Zhou. 2016. Computing tensor eigenvalues via homotopy methods. *SIAM Journal on Matrix Analysis and Applications* 37 (1): 290–319.

61. Kolda, Tamara G., and Jackson R Mayo. 2011. Shifted power method for computing tensor eigenpairs. *SIAM Journal on Matrix Analysis and Applications* 32 (4): 1095–1124.

62. Hitchcock, Frank L. 1927. The expression of a tensor or a polyadic as a sum of products. *Journal of Mathematics and Physics* 6 (1–4): 164–189.

63. Hitchcock, Frank L. 1928. Multiple invariants and generalized rank of a p-way matrix or tensor. *Journal of Mathematics and Physics* 7 (1–4): 39–79.

64. Tucker, Ledyard R. 1963. Implications of factor analysis of three-way matrices for measurement of change. *Problems in Measuring Change* 15 (122–137): 3.

65. Tucker, Ledyard R., et al. 1964. The extension of factor analysis to three-dimensional matrices. *Contributions to Mathematical Psychology* 110119.

66. Tucker, Ledyard R. 1966. Some mathematical notes on three-mode factor analysis. *Psychometrika* 31 (3): 279–311.

67. De Lathauwer, Lieven, Bart De Moor, and Joos Vandewalle. 2000. A multilinear singular value decomposition. *SIAM Journal on Matrix Analysis and Applications* 21 (4): 1253–1278.

68. Bergqvist, Göran, and Erik G Larsson. 2010. The higher-order singular value decomposition: Theory and an application [lecture notes]. *IEEE Signal Processing Magazine* 27 (3): 151–154.

69. Chen, Can, Amit Surana, Anthony Bloch, and Indika Rajapakse. 2019. Data-driven model reduction for multilinear control systems via tensor trains. arXiv:1912.03569.
70. Behera, Ratikanta, and Debasisha Mishra. 2017. Further results on generalized inverses of tensors via the einstein product. *Linear and Multilinear Algebra* 65 (8): 1662–1682.
71. Panigrahy, Krushnachandra, and Debasisha Mishra. 2022. Extension of moore-penrose inverse of tensor via einstein product. *Linear and Multilinear Algebra* 70 (4): 750–773.
72. Gelß, Patrick. 2017. *The tensor-train format and its applications: Modeling and analysis of chemical reaction networks, catalytic processes, fluid flows, and Brownian dynamics*. Ph.D. thesis.
73. Robeva, Elina. 2016. Orthogonal decomposition of symmetric tensors. *SIAM Journal on Matrix Analysis and Applications* 37 (1): 86–102.
74. Anandkumar, Animashree, Rong Ge, Daniel Hsu, Sham M Kakade, and Matus Telgarsky. 2014. Tensor decompositions for learning latent variable models. *Journal of Machine Learning Research* 15: 2773–2832.
75. Klus, Stefan, Patrick Gelß, Sebastian Peitz, and Christof Schütte. 2018. Tensor-based dynamic mode decomposition. *Nonlinearity* 31 (7): 3359.
76. Oseledets, I. V. 2009. Approximation of matrices with logarithmic number of parameters. In *Doklady mathematics*, vol 80, 653–654. Berlin: Springer.
77. Hou, Ming, and Brahim Chaib-Draa. 2015. Hierarchical tucker tensor regression: Application to brain imaging data analysis. In *2015 IEEE international conference on image processing (ICIP)*, 1344–1348. IEEE.
78. Grasedyck, Lars. 2010. Hierarchical singular value decomposition of tensors. *SIAM Journal on Matrix Analysis and Applications* 31 (4): 2029–2054.
79. Lubich, Christian, Thorsten Rohwedder, Reinhold Schneider, and Bart Vandereycken. 2013. Dynamical approximation by hierarchical tucker and tensor-train tensors. *SIAM Journal on Matrix Analysis and Applications* 34 (2): 470–494.

Tucker Product-Based Dynamical Systems

Abstract

The Tucker product-based dynamical system (TPDS) representation was first introduced by Rogers et al. [*Adv. Neural Inf. Process. Syst.*] in 2013. The representation is a generalization of the linear dynamical system (LDS) model which can preserve the state and output of the system as tensors. TPDSs utilize multilinear operators formed by the Tucker product of matrices to govern the system evolution and can be used to capture dynamics with tensor time-series data. Techniques for TPDS model reduction and system identification have been developed. Furthermore, system-theoretic properties, including stability, reachability, and observability, have been extended to TPDSs through tensor algebra, enabling efficient computation.

2.1 Overview

Tensor time-series data represents a unique and powerful way to capture the evolution of complex systems over time [1–7]. Unlike traditional time-series data, which consists of a single sequence of measurements, tensor time-series data embodies a multidimensional array of measurements, enabling the representation of intricate interactions and relationships between multiple variables. This rich data format has emerged as a valuable tool in various fields, including economics [8–10], climatology [11–13], ecology [14–16], and biology [17–19]. For instance, the dynamics of human genome involves the interactions between 3D genome architecture, function, and its relationship to phenotype over time, known as 4D Nucleome [19–21], which naturally fits the tensorial structure.

One prominent application of tensor time-series data lies in forecasting future measurements based on historical observations. This task, known as system identification, aims to uncover the underlying patterns and relationships within the data to predict future trends.

© The Author(s), under exclusive license to Springer Nature Switzerland AG 2024   25
C. Chen, *Tensor-Based Dynamical Systems*,
Synthesis Lectures on Mathematics & Statistics,
https://doi.org/10.1007/978-3-031-54505-4_2

Numerous system identification techniques have been developed for linear and nonlinear vector-based dynamical systems, such as eigensystem realization algorithm [22, 23], balanced proper orthogonal decomposition [24, 25], sparse identification of nonlinear dynamics [26, 27], and others [28–31]. However, these methods cannot be directly applicable to tensor time-series data due to the inherent challenges of handling multidimensional data structures.

A common approach to circumvent this limitation involves vectorizing the tensor time-series data, transforming it into a one-dimensional vector [1, 2]. While this approach enables the application of traditional system identification methods, it leads to an explosion in dimensionality, potentially obscuring the underlying structure of the data and rendering computations intractable. In other words, vectorization disregards the inherent tensorial nature of the data, which can provide valuable information for efficient representation and computation [32].

This chapter discusses the discrete-time Tucker product-based dynamical system (TPDS) proposed in [1]. The evolution of the system is governed by multilinear operators formed using the Tucker product of matrices. The TPDS representation with outputs is defined as

$$\begin{cases} \mathsf{X}_{t+1} = \mathsf{X}_t \times \{\mathbf{A}_1, \mathbf{A}_2, \ldots, \mathbf{A}_k\} \\ \mathsf{Y}_t = \mathsf{X}_t \times \{\mathbf{C}_1, \mathbf{C}_2, \ldots, \mathbf{C}_k\} \end{cases}, \qquad (2.1)$$

where $\mathsf{X}_t \in \mathbb{R}^{n_1 \times n_2 \times \cdots \times n_k}$ are the state variables, $\mathsf{Y}_t \in \mathbb{R}^{m_1 \times m_2 \times \cdots \times m_k}$ are the output variables/measurements, and $\mathbf{A}_p \in \mathbb{R}^{n_p \times n_p}$ and $\mathbf{C}_p \in \mathbb{R}^{m_p \times n_p}$ are real-value parameter matrices for $p = 1, 2, \ldots, k$. The TPDS representation provides a more comprehensive and nuanced understanding of system dynamics, making it particularly well-suited for modeling complex systems with tensor time-series data [1]. After applying the Kronecker product, the TPDS (2.1) can be expressed in a vectorized form as follows:

$$\begin{cases} \mathrm{vec}(\mathsf{X}_{t+1}) = \mathbf{A}\,\mathrm{vec}(\mathsf{X}_t) \\ \mathrm{vec}(\mathsf{Y}_t) = \mathbf{C}\,\mathrm{vec}(\mathsf{X}_t) \end{cases}, \qquad (2.2)$$

where $\mathbf{A} = \mathbf{A}_k \otimes \mathbf{A}_{k-1} \otimes \cdots \otimes \mathbf{A}_1$ and $\mathbf{C} = \mathbf{C}_k \otimes \mathbf{C}_{k-1} \otimes \cdots \otimes \mathbf{C}_1$. Here, the notation "vec" denotes the vectorization operation defined in Chap. 1.

Given tensor time-series measurements $\mathsf{Y}_1, \mathsf{Y}_2, \ldots, \mathsf{Y}_T$, our objective is to estimate the dynamics of the measurements by fitting the TPDS (2.1). The number of parameters in the TPDS (2.1) is only on the order of

$$\mathcal{O}\left(\sum_{p=1}^{k} n_p^2 + n_p m_p \right),$$

while the number of parameters using tensor vectorization is on the order of

$$\mathcal{O}\left(\prod_{p=1}^{k} n_p^2 + \prod_{p=1}^{k} n_p m_p\right).$$

Therefore, the TPDS representation offers significant advantages in terms of parameter efficiency compared to traditional vector-based dynamical systems. Additionally, the framework enables the investigation of fundamental system-theoretic properties, such as stability, reachability, and observability of the TPDS (2.1), once the parameter matrices are obtained. The content of this chapter is mainly based on the work of [1, 2, 32].

2.2 System Identification

This section presents two system identification approaches for TPDSs in the presence of noise, i.e.,

$$\begin{cases} \mathsf{X}_{t+1} = \mathsf{X}_t \times \{\mathbf{A}_1, \mathbf{A}_2, \ldots, \mathbf{A}_k\} + \mathcal{N}_{\text{tensor}}(0, \mathbf{Q}_1, \mathbf{Q}_2, \ldots, \mathbf{Q}_k) \\ \mathsf{Y}_t = \mathsf{X}_t \times \{\mathbf{C}_1, \mathbf{C}_2, \ldots, \mathbf{C}_k\} + \mathcal{N}_{\text{tensor}}(0, \mathbf{R}_1, \mathbf{R}_2, \ldots, \mathbf{R}_k) \end{cases}, \qquad (2.3)$$

where the normal distribution on tensors can be defined as

$$\mathsf{T} \sim \mathcal{N}_{\text{tensor}}(0, \mathbf{Q}_1, \mathbf{Q}_2, \ldots, \mathbf{Q}_k) \Leftrightarrow \text{vec}(\mathsf{T}) \sim \mathcal{N}(\mathbf{0}, \mathbf{Q})$$

with $\mathbf{Q} = \mathbf{Q}_k \otimes \mathbf{Q}_{k-1} \otimes \cdots \otimes \mathbf{Q}_1$. Here, 0 and $\mathbf{0}$ represent the zero tensor and the zero matrix, respectively. The factor matrix $\mathbf{Q}_p$ can be viewed as the p-mode covariance. The goal is to estimate the system parameters $\mathbf{A}_p$, $\mathbf{Q}_p$, $\mathbf{C}_p$, and $\mathbf{R}_p$ using tensor time-series measurements.

2.2.1 EM Algorithm

The expectation maximization (EM) algorithm [33–35] can be applied for estimating the parameters of the TPDS with noises (2.3) [1]. The essence of the approach is to locally maximize the expected complete log-likelihood by computing the gradient with respect to a vector $\mathbf{c} \in \mathbb{R}^{\sum_{p=1}^{k} n_p m_p}$ defined as

$$\mathbf{c} = \left[\text{vec}(\mathbf{C}_1)^{\top} \ \text{vec}(\mathbf{C}_2)^{\top} \ \cdots \ \text{vec}(\mathbf{C}_k)^{\top}\right]^{\top}.$$

The expected complete log-likelihood in terms of $\mathbf{c}$ can be computed as

$$l(\mathbf{c}) = \text{Trace}\left(\mathbf{R}^{-1}\mathbf{C}(\mathbf{\Gamma}\mathbf{C}^{\top} - 2\mathbf{\Omega}^{\top})\right), \qquad (2.4)$$

where $\mathbf{R} = \mathbf{R}_k \otimes \mathbf{R}_{k-1} \otimes \cdots \otimes \mathbf{R}_1$, $\mathbf{C} = \mathbf{C}_k \otimes \mathbf{C}_{k-1} \otimes \cdots \otimes \mathbf{C}_1$, and

$$\boldsymbol{\Gamma} = \sum_{t=1}^{T} \mathbb{E}\Big(\text{vec}(\mathsf{X}_t)\text{vec}(\mathsf{X}_t)^{\top}\Big) \qquad \boldsymbol{\Omega} = \sum_{t=1}^{T} \text{vec}(\mathsf{Y}_t)\mathbb{E}\Big(\text{vec}(\mathsf{X}_t)^{\top}\Big).$$

Here, the notation $\mathbb{E}$ denotes the expectation operation. The element-wise gradient $\nabla l(\mathbf{c})$ then can be computed as

$$\nabla l(\mathbf{c})_q = 2\text{Trace}\Big(\mathbf{R}^{-1}\partial_p\mathbf{C}(\boldsymbol{\Gamma}\mathbf{C}^{\top} - \boldsymbol{\Omega}^{\top})\Big),$$

where $\partial_p\mathbf{C} = \mathbf{C}_k \otimes \mathbf{C}_{k-1} \otimes \cdots \otimes \boldsymbol{\Delta}_p \otimes \cdots \otimes \mathbf{C}_1$. Here, $\boldsymbol{\Delta}_p$ is an indicator matrix that corresponds to the qth entry of $\mathbf{c}$.

The M-step for estimating $\mathbf{A}_p$ can be achieved in a similar manner by considering a concatenated vector $\mathbf{a} \in \mathbb{R}^{\sum_{p=1}^{k} n_p^2}$ defined as

$$\mathbf{a} = \Big[\text{vec}(\mathbf{A}_1)^{\top}\ \text{vec}(\mathbf{A}_2)^{\top}\ \cdots\ \text{vec}(\mathbf{A}_k)^{\top}\Big]^{\top}$$

with the expected complete log-likelihood

$$l(\mathbf{a}) = \text{Trace}\Big(\mathbf{Q}^{-1}\mathbf{A}(\boldsymbol{\Gamma}\mathbf{A}^{\top} - 2\boldsymbol{\Omega}^{\top})\Big), \tag{2.5}$$

where $\mathbf{Q} = \mathbf{Q}_k \otimes \mathbf{Q}_{k-1} \otimes \cdots \otimes \mathbf{Q}_1$, $\mathbf{A} = \mathbf{A}_k \otimes \mathbf{A}_{k-1} \otimes \cdots \otimes \mathbf{A}_1$, and

$$\boldsymbol{\Gamma} = \sum_{t=1}^{T-1} \mathbb{E}\Big(\text{vec}(\mathsf{X}_t)\text{vec}(\mathsf{X}_t)^{\top}\Big) \qquad \boldsymbol{\Omega} = \sum_{t=1}^{T-1} \mathbb{E}\Big(\text{vec}(\mathsf{X}_{t+1})\text{vec}(\mathsf{X}_t)^{\top}\Big).$$

The EM algorithm has proven to be an effective method for estimating the parameters of TPDSs with noise [1]. By alternately updating the expectations of the latent variables and the parameters, this approach circumvents the difficulty of directly maximizing the likelihood of the data with respect to the parameters [1]. However, it is worth noting that this approach does not consider model reduction, which is a critical factor in dealing with high-dimensional TPDSs.

2.2.2 Multilinear PCA/Regression

To tackle the issue of model reduction in system identification for TPDSs driven by noise, Surana et al. [2] utilized multilinear principal component analysis (PCA) and multilinear regression. Given tensor time-series measurements $\{\mathsf{Y}_1, \mathsf{Y}_2, \ldots, \mathsf{Y}_T\}$, the goal of multilinear PCA is to find transformation matrices $\mathbf{U}_p \in \mathbb{R}^{m_p \times n_p}$ with $n_p < m_p$ such that

$$\mathsf{X}_t = \mathsf{Y}_t \times \{\mathbf{U}_1^{\top}, \mathbf{U}_2^{\top}, \ldots, \mathbf{U}_k^{\top}\}, \tag{2.6}$$

and the scatter of the low-dimensional latent variables $\sum_{t=1}^{T} \|X_t - \bar{X}\|^2$ is minimized (where $\bar{X}$ denotes the mean of X_t). Multilinear PCA can be solved by iterative approaches, see detailed algorithms in [2].

Given a set of input and output tensors $\{(X_t, Y_t)\}_{t=1}^{T}$ with $X_t \in \mathbb{R}^{n_1 \times n_2 \times \cdots \times n_k}$ and $Y_t \in \mathbb{R}^{m_1 \times m_2 \times \cdots \times m_k}$, the objective of multilinear regression is to fit the Tucker product-based model, i.e.,

$$Y_t = X_t \times \{\mathbf{M}_1, \mathbf{M}_2, \ldots, \mathbf{M}_k\} + \mathcal{N}_{\text{tensor}}(0, \mathbf{R}_1, \mathbf{R}_2, \ldots, \mathbf{R}_k) \tag{2.7}$$

for $\mathbf{M}_p \in \mathbb{R}^{m_p \times n_p}$ and $\mathbf{R}_p \in \mathbb{R}^{m_p \times m_p}$. Multilinear regression can also be solved by iterative approaches, see detailed algorithms in [2]. In particular, the noise covariance matrices $\mathbf{R}_p$ can be solved from the technique of maximum likelihood estimation after finding $\mathbf{M}_p$.

The system identification of the TPDS (2.3) with model reduction can be achieved through the following steps:

- First, compute multilinear PCA of $\{Y_t\}_{t=1}^{T}$ to obtain the latent variables $\{X_t\}_{t=1}^{T}$ with the factor matrices $\{\mathbf{C}_1, \mathbf{C}_2, \ldots, \mathbf{C}_k\}$;
- Subsequently, use the maximum likelihood estimation to obtain the noise covariance matrices $\{\mathbf{R}_1, \mathbf{R}_2, \ldots, \mathbf{R}_k\}$ based on the residuals of Y_t;
- Finally, apply multilinear regression to $\{(X_t, X_{t+1})\}_{t=1}^{T-1}$ to obtain the factor matrices $\{\mathbf{A}_1, \mathbf{A}_2, \ldots, \mathbf{A}_k\}$ and the noise covariance matrices $\{\mathbf{Q}_1, \mathbf{Q}_2, \ldots, \mathbf{Q}_k\}$.

Multilinear PCA and multilinear regression offer an effective approach to identify the parameters of TPDSs driven by noise while preserving low-dimensional latent variables [2]. After obtaining the system parameter matrices, i.e., $\mathbf{A}_p$ and $\mathbf{C}_p$, it is significant to study the system-theoretic properties of TPDSs.

2.3 System-Theoretic Properties

This section explores the notions of stability, reachability, and observability for TPDSs. While these notions can be determined using tensor vectorization, the Tucker product representation provides a more efficient approach to compute them.

2.3.1 Stability

The stability of the TPDS (2.1, top) can be defined analogously to that of LDSs. Analyzing the stability properties of the equilibrium point $X_e = 0$ of the TPDS is of particular interest.

Definition 2.1 ([32]) The equilibrium point $X_e = 0$ of the TPDS (2.1, top) is called:

- stable if $\|X_t\| \leq \gamma \|X_0\|$ for some $\gamma > 0$;
- asymptotically stable if $\lim_{t\to\infty} \|X_t\| = 0$;
- unstable if it is not stable.

The stability of the TPDS (2.1, top) can be efficiently determined by examining the eigenvalues of the factor matrices $\mathbf{A}_p$.

Proposition 2.1 ([32]) *Given the TPDS (2.1, top), the equilibrium point $X_e = 0$ is:*

- *stable if and only if $|\lambda_{j_1}^{(1)}\lambda_{j_2}^{(2)} \cdots \lambda_{j_k}^{(k)}| \leq 1$ for all $j_p = 1, 2, \ldots, n_p$ (and if $|\lambda_{j_1}^{(1)}\lambda_{j_2}^{(2)} \cdots \lambda_{j_k}^{(k)}| = 1$, all the eigenvalues must have equal algebraic and geometric multiplicity);*
- *asymptotically stable if and only if $|\lambda_{j_1}^{(1)}\lambda_{j_2}^{(2)} \cdots \lambda_{j_k}^{(k)}| < 1$ for all $j_p = 1, 2, \ldots, n_p$;*
- *unstable if and only if $|\lambda_{j_1}^{(1)}\lambda_{j_2}^{(2)} \cdots \lambda_{j_k}^{(k)}| > 1$ for some $j_p = 1, 2, \ldots, n_p$,*

where $\lambda_{j_p}^{(p)}$ are the eigenvalues of $\mathbf{A}_p$ for $j_p = 1, 2, \ldots, n_p$ and $p = 1, 2, \ldots, k$.

Proof As mentioned, the dynamic matrix of the vectorized representation of the TPDS is $\mathbf{A} = \mathbf{A}_k \otimes \mathbf{A}_{k-1} \otimes \cdots \otimes \mathbf{A}_1$. Based on the fact that the eigenvalues of $\mathbf{A}$ are equal to the products of the eigenvalues of its factor matrices (note that the eigenvalue of $\mathbf{A}$ has equal algebraic and geometric multiplicity if its corresponding eigenvalues from the factor matrices have equal algebraic and geometric multiplicity), the stability of TPDSs can be efficiently determined using linear stability analysis. $\qquad\square$

The results can be further simplified by considering the spectral radii, which are the maximum absolute eigenvalues, of the factor matrices $\mathbf{A}_p$.

Corollary 2.1 ([32]) *Given the TPDS (2.1, top), the equilibrium point $X_e = 0$ is:*

- *stable if and only if $\lambda_{sr}^{(1)}\lambda_{sr}^{(2)} \cdots \lambda_{sr}^{(k)} \leq 1$ (and if $\lambda_{sr}^{(1)}\lambda_{sr}^{(2)} \cdots \lambda_{sr}^{(k)} = 1$, all the eigenvalues must have equal algebraic and geometric multiplicity);*
- *asymptotically stable if and only if $\lambda_{sr}^{(1)}\lambda_{sr}^{(2)} \cdots \lambda_{sr}^{(k)} < 1$;*
- *unstable if and only if $\lambda_{sr}^{(1)}\lambda_{sr}^{(2)} \cdots \lambda_{sr}^{(k)} > 1$,*

where $\lambda_{sr}^{(p)}$ are the spectral radii of $\mathbf{A}_p$ for $p = 1, 2, \ldots, k$.

Proof The results can be directly derived from Proposition 2.1. $\qquad\square$

Remark 2.1 The computation of eigenvalues of the factor matrices has a complexity of approximately $\mathcal{O}(\sum_{p=1}^{k} n_p^3)$, while computing the eigenvalues of the dynamic matrix from the vectorized representation requires a complexity of approximately $\mathcal{O}(\prod_{p=1}^{k} n_p^3)$ operations.

2.3.2 Reachability

The discrete-time TPDS (2.1) can be extended by incorporating control inputs, i.e.,

$$\begin{cases} \mathsf{X}_{t+1} = \mathsf{X}_t \times \{\mathbf{A}_1, \mathbf{A}_2, \ldots, \mathbf{A}_k\} + \mathsf{U}_t \times \{\mathbf{B}_1, \mathbf{B}_2, \ldots, \mathbf{B}_k\} \\ \mathsf{Y}_t = \mathsf{X}_t \times \{\mathbf{C}_1, \mathbf{C}_2, \ldots, \mathbf{C}_k\} \end{cases}, \qquad (2.8)$$

where $\mathsf{U}_t \in \mathbb{R}^{s_1 \times s_2 \times \cdots \times s_k}$ are the control inputs, and $\mathbf{B}_p \in \mathbb{R}^{n_p \times s_p}$ are real-value control matrices. Similar to LDSs, the reachability of the TPDS (2.8) can be defined as follows.

Definition 2.2 ([32]) The TPDS (2.8, top) is said to be reachable on the interval $[t_0, t_f]$ if for any two states $\mathsf{X}_{t_0} = \mathsf{X}_0$ and $\mathsf{X}_{t_f} = \mathsf{X}_f$ there exists a sequence of control inputs U_t that drive the system from X_0 to X_f.

Proposition 2.2 *The TPDS (2.8, top) is reachable over the interval $[t_0, t_f]$ if and only if the reachability Gramian defined as*

$$\boldsymbol{G}_R = \sum_{t=t_0}^{t_f-1} \boldsymbol{G}_k \otimes \boldsymbol{G}_{k-1} \otimes \cdots \otimes \boldsymbol{G}_1, \qquad (2.9)$$

where $\boldsymbol{G}_p = \boldsymbol{A}_p^{t_f-t-1} \boldsymbol{B}_p \boldsymbol{B}_p^{\top} (\boldsymbol{A}_p^{\top})^{t_f-t-1}$ for $p = 1, 2, \ldots, k$, is positive definite.

Proof The dynamic matrix and control matrix of the vectorized representation of the TPDS are $\mathbf{A} = \mathbf{A}_k \otimes \mathbf{A}_{k-1} \otimes \cdots \otimes \mathbf{A}_1$ and $\mathbf{B} = \mathbf{B}_k \otimes \mathbf{B}_{k-1} \otimes \cdots \otimes \mathbf{B}_1$, respectively. By using the properties of the Kronecker product, the reachability Gramian of the TPDS can be expressed as

$$\begin{aligned} \boldsymbol{G}_R &= \sum_{t=t_0}^{t_f-1} \mathbf{A}^{t_f-t-1} \mathbf{B}\mathbf{B}^{\top} (\mathbf{A}^{\top})^{t_f-t-1} \\ &= \sum_{t=t_0}^{t_f-1} (\mathbf{A}_k \otimes \mathbf{A}_{k-1} \otimes \cdots \otimes \mathbf{A}_1)^{t_f-t-1} (\mathbf{B}_k \otimes \mathbf{B}_{k-1} \otimes \cdots \otimes \mathbf{B}_1) \\ &\qquad (\mathbf{B}_k \otimes \mathbf{B}_{k-1} \otimes \cdots \otimes \mathbf{B}_1)^{\top} ((\mathbf{A}_k \otimes \mathbf{A}_{k-1} \otimes \cdots \otimes \mathbf{A}_1)^{\top})^{t_f-t-1} \\ &= \sum_{t=t_0}^{t_f-1} \boldsymbol{G}_k \otimes \boldsymbol{G}_{k-1} \otimes \cdots \otimes \boldsymbol{G}_1 \end{aligned}$$

for $\mathbf{G}_p = \mathbf{A}_p^{t_f-t-1}\mathbf{B}_p\mathbf{B}_p^\top(\mathbf{A}_p^\top)^{t_f-t-1}$. Therefore, the results follow immediately. $\square$

Similar to LDSs, a Kalman's rank condition can also be derived for TPDSs.

Proposition 2.3 *The TPDS (2.8, top) is reachable if and only if the reachability matrix defined as*

$$R = \begin{bmatrix} R_0 \; R_1 \; \cdots \; R_{\prod_{p=1}^k n_p - 1} \end{bmatrix}, \tag{2.10}$$

where $R_j = (A_k^j B_k) \otimes (A_{k-1}^j B_{k-1}) \otimes \cdots \otimes (A_1^j B_1)$, *has full rank.*

Proof The proof is similar to Proposition 2.2. $\square$

Remark 2.2 The time complexity of computing the reachability matrix (2.10) is approximately given by

$$\mathcal{O}\left(\prod_{p=1}^k n_p^2 s_p + \sum_{j=1}^{\prod_{p=1}^k n_p - 1} \sum_{p=1}^k n_p^j s_p \right),$$

which is much lower than the time complexity of computing the reachability matrix from the vectorized representation, i.e.,

$$\mathcal{O}\left(\sum_{j=1}^{\prod_{p=1}^k n_p - 1} \prod_{p=1}^k n_p^j s_p \right).$$

2.3.3 Observability

The definitions of observability, observability Gramians, and observability matrices can be defined similarly using the duality principle.

Definition 2.3 ([32]) The TPDS (2.8) is said to be observable on an interval $[t_0, t_f]$ if any initial state $\mathsf{X}_{t_0} = \mathsf{X}_0$ can be uniquely determined from the outputs Y_t.

Proposition 2.4 *The TPDS (2.8) is observable over the interval $[t_0, t_f]$ if and only if the observability Gramian defined as*

$$G_O = \sum_{t=t_0}^{t_f-1} G_k \otimes G_{k-1} \otimes \cdots \otimes G_1, \tag{2.11}$$

where $G_p = (A_p^\top)^{t-t_0} C_p^\top C_p A_p^{t-t_0}$ *for* $p = 1, 2, \ldots, k$, *is positive definite.*

Proposition 2.5 *The TPDS (2.8) is observable if and only if the observability matrix defined as*

$$O = \begin{bmatrix} O_0 \ O_1 \ \cdots \ O_{\prod_{p=1}^k n_p - 1} \end{bmatrix}^\top, \tag{2.12}$$

where $O_j = (C_k A_k^j) \otimes (C_{k-1} A_{k-1}^j) \otimes \cdots \otimes (C_1 A_1^j)$, has full rank.

2.4 Applications

2.4.1 Video Dataset

This example, adapted from [2], aims to demonstrate system identification of TPDSs driven by noise using multilinear PCA/regression. Three dynamic texture videos, traffic, waterfall, and flame, with a total of 52, 150, and 200 time points, respectively, are provided, as shown in Fig. 2.1A, B, and C. The time-series measurements of the videos are third-order tensors, since there are three channels. 80% of the time-series measurements are used for model fitting, while the rest serve for prediction purposes. Simultaneously, the three videos are also fitted by the LDS model with PCA, with 80% training and 20% testing. The results of the relative error defined as

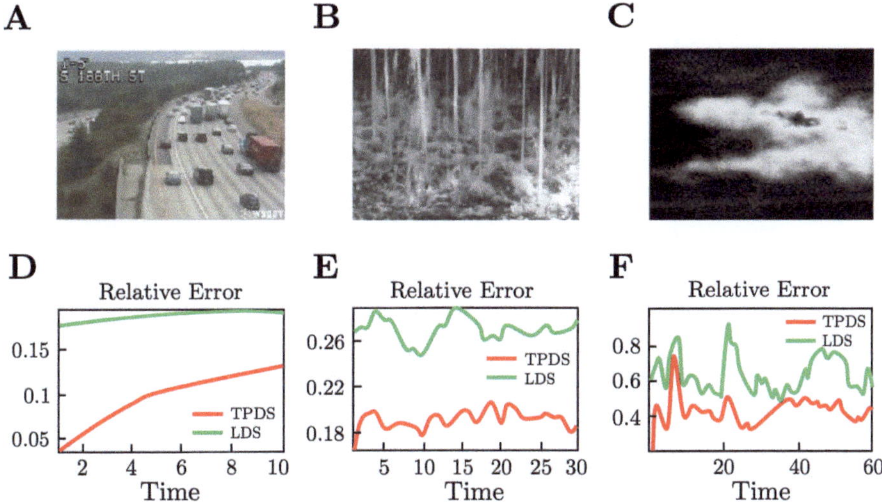

Fig. 2.1 **A**, **B**, and **C** represent snapshots of the three dynamic texture videos—traffic, waterfall, and flame, respectively. **D**, **E**, and **F** show the relative error comparisons between TPDS and LDS for the three videos. This figure was redrawn from [2] with permission

$$\text{Relative Error} = \frac{\|\mathsf{Y}_t - \hat{\mathsf{Y}}_t\|}{\|\mathsf{Y}_t\|},$$

where Y_t and $\hat{\mathsf{Y}}_t$ are the true and predicted measurements at time t, for the two models are shown in Fig. 2.1D, E, and F. It is clear that using TPDSs to fit the measurements achieves lower relative errors compared to using the vectorized LDS with low memory consumption.

2.4.2 Synthetic Data: Stability

This example considers a TPDS defined by (2.1, top), with factor matrices given as

$$\mathbf{A}_1 = \begin{bmatrix} 0 & 0.8 & 0 \\ 0.2 & 0 & 0 \\ 0.5 & 0 & 1 \end{bmatrix}, \quad \mathbf{A}_2 = \begin{bmatrix} 1 & 0.8 \\ 0 & 0.5 \end{bmatrix}, \quad \mathbf{A}_3 = \begin{bmatrix} 0.9 & 1 \\ 0 & 0.5 \end{bmatrix}.$$

The spectral radii of the three factor matrices are 1, 1, and 0.9, respectively. Based on Corollary 2.1, it can be implied that the TPDS is asymptotically stable. Refer to Fig. 2.2 for numerical simulations.

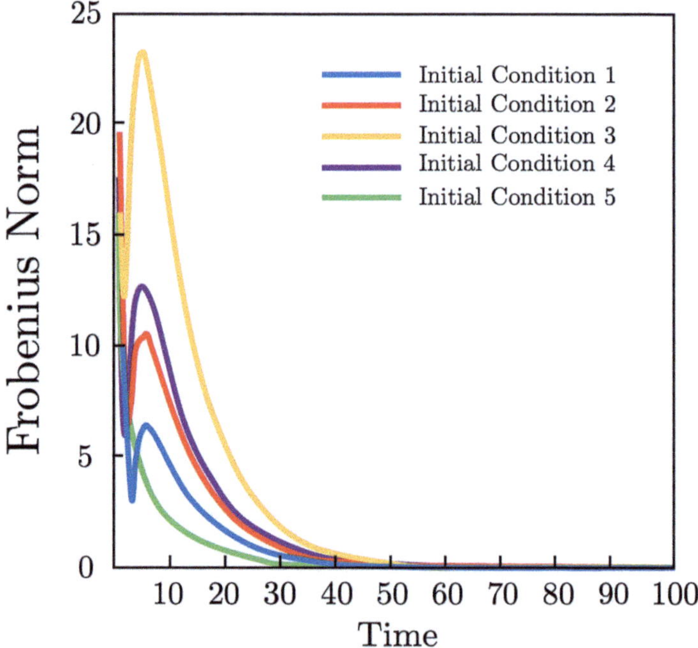

Fig. 2.2 Trajectories of the TPDS with five random initial conditions

2.4.3 Synthetic Data: Reachability and Observability

This example, adapted from [32], examines a TPDS of the form (2.8) with factor matrices

$$\mathbf{A}_1 = \begin{bmatrix} 0 & 1 & 0 \\ 0 & 0 & 1 \\ 0.2 & 0.5 & 0.8 \end{bmatrix}, \quad \mathbf{A}_2 = \begin{bmatrix} 0 & 1 \\ 0.5 & 0, \end{bmatrix},$$

$$\mathbf{B}_1 = \begin{bmatrix} 0 \\ 0 \\ 1 \end{bmatrix}, \quad \mathbf{B}_2 = \begin{bmatrix} 0 \\ 1 \end{bmatrix},$$

$$\mathbf{C}_1 = \begin{bmatrix} 1 & 0 & 0 \end{bmatrix}, \quad \mathbf{C}_2 = \begin{bmatrix} 1 & 0 \end{bmatrix}.$$

The state variable $\mathsf{X}_t \in \mathbb{R}^{3 \times 2}$ are second-order tensors, i.e., matrices. According to Propositions 2.3 and 2.5, the reachability and observability matrices can be computed as

$$\mathbf{R} = \begin{bmatrix} 0 & 0 & 0 & 0.4 & 0 & 0.378 \\ 0 & 1 & 0 & 0.57 & 0 & 0.4849 \\ 0 & 0.8 & 0 & 0.756 & 0 & 0.6339 \\ 0 & 0 & 0.5 & 0 & 0.285 & 0 \\ 0 & 0 & 0.4 & 0 & 0.378 & 0 \\ 1 & 0 & 0.57 & 0 & 0.4849 & 0 \end{bmatrix}$$

and

$$\mathbf{O} = \begin{bmatrix} 1 & 0 & 0 & 0 & 0 & 0 \\ 0 & 0 & 0 & 0 & 1 & 0 \\ 0 & 0 & 0.5 & 0 & 0 & 0 \\ 0 & 0 & 0 & 0.1 & 0.25 & 0.4 \\ 0.04 & 0.15 & 0.285 & 0 & 0 & 0 \\ 0 & 0 & 0 & 0.057 & 0.1825 & 0.378 \end{bmatrix},$$

respectively. The two matrices have full rank, so the TPDS is both reachable and observable.

References

1. Rogers, Mark, Lei Li, and Stuart J Russell. 2013. Multilinear dynamical systems for tensor time series. *Advances in Neural Information Processing Systems* 26.
2. Surana, Amit, Geoff Patterson, and Indika Rajapakse. 2016. Dynamic tensor time series modeling and analysis. In *2016 IEEE 55th conference on decision and control (CDC)*, 1637–1642. IEEE.
3. Liang, Paul Pu, Zhun Liu, Yao-Hung Hubert Tsai, Qibin Zhao, Ruslan Salakhutdinov, and Louis-Philippe Morency. 2019. Learning representations from imperfect time series data via tensor rank regularization. arXiv:1907.01011.
4. Foster, Grant. 1996. Time series analysis by projection. ii. tensor methods for time series analysis. *The Astronomical Journal* 111: 555.

5. Jing, Baoyu, Hanghang Tong, and Yada Zhu. 2021. Network of tensor time series. *In Proceedings of the Web Conference* 2021: 2425–2437.
6. Chen, Rong, Dan Yang, and Cun-Hui. Zhang. 2022. Factor models for high-dimensional tensor time series. *Journal of the American Statistical Association* 117 (537): 94–116.
7. Chen, Xinyu, and Lijun Sun. 2021. Bayesian temporal factorization for multidimensional time series prediction. *IEEE Transactions on Pattern Analysis and Machine Intelligence.*
8. Chen, Weilin, and Clifford Lam. 2022. Rank and factor loadings estimation in time series tensor factor model by pre-averaging. arXiv:2208.04012.
9. Han, Yuefeng, Rong Chen, and Cun-Hui. Zhang. 2022. Rank determination in tensor factor model. *Electronic Journal of Statistics* 16 (1): 1726–1803.
10. Beck, Nathaniel, and Jonathan N Katz. 2011. Modeling dynamics in time-series–cross-section political economy data. *Annual Review of Political Science* 14: 331–352.
11. Bahadori, Mohammad Taha, Qi Rose Yu, and Yan Liu. 2014. Fast multivariate spatio-temporal analysis via low rank tensor learning. In *Advances in Neural information processing systems*, 27.
12. Jing, Peiguang, Su. Yuting, Xiao Jin, and Chengqian Zhang. 2018. High-order temporal correlation model learning for time-series prediction. *IEEE Transactions on Cybernetics* 49 (6): 2385–2397.
13. Yu, Hsiang-Fu, Nikhil Rao, and Inderjit S Dhillon. 2016. Temporal regularized matrix factorization for high-dimensional time series prediction. In *Advances in neural information processing systems*, 29.
14. Frelat, Romain, Martin Lindegren, Tim Spaanheden Denker, Jens Floeter, Heino O Fock, Camilla Sguotti, Moritz Stäbler, Saskia A Otto, and Christian Möllmann. 2017. Community ecology in 3d: Tensor decomposition reveals spatio-temporal dynamics of large ecological communities. *PloS One* 12 (11): e0188205.
15. Korevaar, Hannah, C. Jessica Metcalf, and Bryan T. Grenfell. 2020. Tensor decomposition for infectious disease incidence data. *Methods in Ecology and Evolution* 11 (12): 1690–1700.
16. Zhang, Xueqi, Meng Zhao, and Rencai Dong. 2020. Time-series prediction of environmental noise for urban iot based on long short-term memory recurrent neural network. *Applied Sciences* 10 (3): 1144.
17. Yahyanejad, Farzane, Réka. Albert, and Bhaskar DasGupta. 2019. A survey of some tensor analysis techniques for biological systems. *Quantitative Biology* 7 (4): 266–277.
18. Skantze, Viktor, Mikael Wallman, Ann-Sofie. Sandberg, Rikard Landberg, Mats Jirstrand, and Carl Brunius. 2023. Identification of metabotypes in complex biological data using tensor decomposition. *Chemometrics and Intelligent Laboratory Systems* 233: 104733.
19. Dekker, Job, Andrew S Belmont, Mitchell Guttman, Victor O Leshyk, John T Lis, Stavros Lomvardas, Leonid A Mirny, Clodagh C O'shea, Peter J Park, Bing Ren, et al. 2017. The 4d nucleome project. *Nature* 549 (7671): 219–226.
20. Chen, Can, Amit Surana, Anthony Bloch, and Indika Rajapakse. 2019. Multilinear time invariant system theory. In *2019 proceedings of the conference on control and its applications*, 118–125. SIAM.
21. Chen, Haiming, Jie Chen, Lindsey A Muir, Scott Ronquist, Walter Meixner, Mats Ljungman, Thomas Ried, Stephen Smale, and Indika Rajapakse. 2015. Functional organization of the human 4d nucleome. *Proceedings of the National Academy of Sciences* 112 (26): 8002–8007.
22. Juang, Jer-Nan, and Richard S Pappa. 1985. An eigensystem realization algorithm for modal parameter identification and model reduction. *Journal of Guidance, Control, and Dynamics* 8 (5): 620–627.
23. Pappa, Richard S, Kenny B Elliott, and Axel Schenk. 1993. Consistent-mode indicator for the eigensystem realization algorithm. *Journal of Guidance, Control, and Dynamics* 16 (5): 852–858.
24. Rowley, Clarence W. 2005. Model reduction for fluids, using balanced proper orthogonal decomposition. *International Journal of Bifurcation and Chaos* 15 (03): 997–1013.

25. Ilak, Miloš, and Clarence W Rowley. 2008. Modeling of transitional channel flow using balanced proper orthogonal decomposition. *Physics of Fluids* 20 (3): 034103.

26. Brunton, Steven L, Joshua L Proctor, and J Nathan Kutz. 2016. Discovering governing equations from data by sparse identification of nonlinear dynamical systems. *Proceedings of the National Academy of Sciences* 113 (15): 3932–3937.

27. Kaiser, Eurika, J Nathan Kutz, and Steven L Brunton. 2018. Sparse identification of nonlinear dynamics for model predictive control in the low-data limit. *Proceedings of the Royal Society A* 474 (2219): 20180335.

28. Voss, Henning U, Jens Timmer, and Jürgen Kurths. 2004. Nonlinear dynamical system identification from uncertain and indirect measurements. *International Journal of Bifurcation and Chaos* 14 (06): 1905–1933.

29. Kuschewski, John G, Stefen Hui, and Stanislaw H Zak. 1993. Application of feedforward neural networks to dynamical system identification and control. *IEEE Transactions on Control Systems Technology* 1 (1): 37–49.

30. Green, Peter L. 2015. Bayesian system identification of a nonlinear dynamical system using a novel variant of simulated annealing. *Mechanical Systems and Signal Processing* 52: 133–146.

31. Li, Shou-Ju., and Ying-Xi. Liu. 2006. An improved approach to nonlinear dynamical system identification using pid neural networks. *International Journal of Nonlinear Sciences and Numerical Simulation* 7 (2): 177–182.

32. Chen, Can, Amit Surana, Anthony M Bloch, and Indika Rajapakse. 2021. Multilinear control systems theory. *SIAM Journal on Control and Optimization* 59 (1): 749–776.

33. Moon, Todd K. 1996. The expectation-maximization algorithm. *IEEE Signal Processing Magazine* 13 (6): 47–60.

34. Dellaert, Frank. 2002. *The expectation maximization algorithm*. Georgia Institute of Technology: Technical report.

35. Do, Chuong B., and Serafim Batzoglou. 2008. What is the expectation maximization algorithm? *Nature Biotechnology* 26 (8): 897–899.

Einstein Product-Based Dynamical Systems

3

Abstract

The Einstein product-based dynamical system (EPDS) representation was initially proposed by Chen et al. [*Proc. Conf. Control Appl.*] in 2019, which utilizes an even-order tensor equipped with the Einstein product to drive the system evolution. This representation is an extension of the TPDS representation and has a form similar to the standard LDS model. EPDSs are well-suited for exploring system-theoretic properties, including stability, reachability, and observability, as well as for developing methods for model reduction and system identification. More importantly, tensor decomposition techniques, such as HOSVD, CPD, and TTD, can be exploited on the even-order dynamic tensors in order to facilitate computation and memory efficiency for EPDSs.

3.1 Overview

The Tucker product-based dynamical system (TPDS) representation (2.1) or (2.8) has proven to be an effective tool for modeling and analyzing tensor time-series data, offering a novel framework to capture the complex dynamics of multidimensional systems [1, 2]. However, it is limited by its assumption that the multilinear operators are constructed solely using the Tucker product of matrices [3]. This limitation restricts the expressiveness of the TPDS representation and may not adequately represent intricate relationships between variables within the data. To overcome this limitation and capture more general relationships between variables, a new representation is introduced using even-order tensors and the Einstein product [4]. The extension, referred to as the Einstein product-based dynamical system (EPDS) representation, bears similarities to the LDS model [3, 4].

C. Chen, *Tensor-Based Dynamical Systems*,
Synthesis Lectures on Mathematics & Statistics,
https://doi.org/10.1007/978-3-031-54505-4_3

The discrete-time EPDS with control inputs and outputs is defined as

$$
\begin{cases}
\mathsf{X}_{t+1} = \mathsf{A} * \mathsf{X}_t + \mathsf{B} * \mathsf{U}_t \\
\mathsf{Y}_t = \mathsf{C} * \mathsf{X}_t
\end{cases}, \tag{3.1}
$$

where $\mathsf{X}_t \in \mathbb{R}^{n_1 \times n_2 \times \cdots \times n_k}$ is the state variable, $\mathsf{Y}_t \in \mathbb{R}^{m_1 \times m_2 \times \cdots \times m_k}$ is the output variable, $\mathsf{U}_t \in \mathbb{R}^{s_1 \times s_2 \times \cdots \times s_k}$ is the control input, and $\mathsf{A} \in \mathbb{R}^{n_1 \times n_1 \times \cdots \times n_k \times n_k}$, $\mathsf{B} \in \mathbb{R}^{n_1 \times s_1 \times \cdots \times n_k \times s_k}$, and $\mathsf{C} \in \mathbb{R}^{m_1 \times n_1 \times \cdots \times m_k \times n_k}$ are even-order parameter tensors. Compared to the tensor-based LDS model discussed in [5], the EPDS representation (3.1) is more concise and systematic. Crucially, the EPDS representation provides distinct advantages over the TPDS representation in analyzing system-theoretic properties and tensor time-series data.

First of all, the EPDS (3.1) is indeed a generalization of TPDS (2.8). By taking the outer products of the component matrices $\{\mathbf{A}_1, \mathbf{A}_2, \ldots, \mathbf{A}_k\}$, $\{\mathbf{B}_1, \mathbf{B}_2, \ldots, \mathbf{B}_k\}$, and $\{\mathbf{C}_1, \mathbf{C}_2, \ldots, \mathbf{C}_k\}$ from the TPDS (2.8), it can be rewritten in the form of (3.1). However, the converse of the transformation is not always true. It holds only when the generalized CP ranks of A, B, and C all are equal to one. Additionally, the EPDS (3.1) has a vectorization form, i.e.,

$$
\begin{cases}
\mathrm{vec}(\mathsf{X}_{t+1}) = \psi(\mathsf{A})\mathrm{vec}(\mathsf{X}_t) + \psi(\mathsf{B})\mathrm{vec}(\mathsf{U}_t) \\
\mathrm{vec}(\mathsf{Y}_t) = \psi(\mathsf{C})\mathrm{vec}(\mathsf{X}_t)
\end{cases}, \tag{3.2}
$$

where ψ is the tensor unfolding defined in (1.14). The unfolding-based formulation (3.2) serves as a powerful tool for analyzing EPDSs, enabling the straightforward derivation of system-theoretic properties and the development of model reduction and system identification techniques.

Ultimately, to fully harness the computational advantages of EPDSs, it is essential to work with the tensor form, which facilitates efficient representation and computation of hidden patterns. This entails applying tensor decompositions, such as HOSVD, CPD, and TTD, to the even-order dynamic tensor A, control tensor B, and output tensor C of the EPDS (3.1). These decompositions can substantially reduce the computational and memory burden associated with determining stability, reachability, and observability, as well as model reduction and system identification by leveraging the inherent low-rank structures within the tensors. In particular, numerous tensor algebra operations, such as the Einstein product, TSVD, or solving multilinear systems, can be performed efficiently in the tensor decomposition format compared to matrix methods based on tensor unfoldings [3, 6–8].

This chapter delves into an exploration of the system-theoretic properties, encompassing stability, reachability, and observability, of the EPDS (3.1) through the lens of various tensor decompositions. Furthermore, the chapter discusses how TTD can be employed for model reduction and system identification of EPDSs. The content of this chapter is primarily based on the work of [3, 4, 8].

3.2 System-Theoretic Properties

The definitions of stability, reachability, and observability of the EPDS (3.1) are identical to those defined for TPDSs in Chap. 2.

3.2.1 Stability

The stability of the unforced ETDS (3.1), i.e.,

$$\mathsf{X}_{t+1} = \mathsf{A} * \mathsf{X}_t \tag{3.3}$$

can be obtained by exploiting the U-eigenvalues of A.

Proposition 3.1 ([3, 4]) *Given the EPDS (3.3), the equilibrium point* $\mathsf{X}_e = 0$ *is*

- *stable if and only if* $|\lambda_j| \leq 1$ *for all* $j = 1, 2, \ldots, \prod_{p=1}^{k} n_p$ *(and if* $|\lambda_j| = 1$*, these U-eigenvalues must have equal algebraic and geometry multiplicities);*
- *asymptotically stable if and only if* $|\lambda_j| < 1$ *for all* $j = 1, 2, \ldots, \prod_{p=1}^{k} n_p$*;*
- *unstable if and only if* $|\lambda_j| > 1$ *for some* $j = 1, 2, \ldots, \prod_{p=1}^{k} n_p$*.*

Proof The proof is straightforward and relies on the vectorization form of (3.3), which is $\mathrm{vec}(\mathsf{X}_{k+1}) = \psi(\mathsf{A})\mathrm{vec}(\mathsf{X}_k)$. By using the definition of U-eigenvalues and the linear stability conditions, the result follows immediately. It should be noted that the multiplicity of U-eigenvalues can also be defined similarly via the tensor unfolding ψ. $\qquad\square$

Computing the U-eigenvalues of A involves tensor unfolding and matrix eigenvalue decomposition and requires a computational cost of $\mathcal{O}(\prod_{p=1}^{k} n_p^3)$ operations, making it computationally expensive for large EPDSs. To address this issue, the following results establish connections between U-eigenvalues and HOSVD, CPD, and TTD, which can provide computational advantages in computing or approximating the U-eigenvalues of even-order tensors.

Corollary 3.1 ([3]) *Suppose that* A *is given in the HOSVD form. The equilibrium point* $\mathsf{X}_e = 0$ *of the EPDS (3.3) is asymptotically stable if the sum of the p-mode singular values squared is less than one for any p, i.e.,*

$$\sum_{j_p=1}^{n_p} (\gamma_{j_p}^{(p)})^2 < 1.$$

Proof The sum of the p-mode singular values squared is equal to the Frobenius norm of A squared, as shown by De Lathauwer et al. [9]. Therefore, it follows that

$$\sum_{j_p=1}^{n_p} (\gamma_{j_p}^{(p)})^2 = \|A\|^2 = \|\psi(A)\|^2$$

for any p. According to linear algebra, the spectral radius of a matrix is always less than or equal to its Frobenius norm. Hence, the magnitude of the maximal U-eigenvalue (or U-spectral radius) satisfies

$$\lambda_{sr} \leq \sum_{j_p=1}^{n_p} (\gamma_{j_p}^{(p)})^2$$

for any p, and the result follows from Proposition 3.1. $\qquad\qquad\square$

Lemma 3.1 ([3]) *Suppose that $T \in \mathbb{R}^{n_1 \times m_1 \times \cdots \times n_k \times m_k}$ is given in the CPD form, where at least two of its factor matrices $V^{(p)}$ and $V^{(q)}$ have all column vectors orthonormal for odd p and even q. The matrix SVD of $\psi(T)$ can be obtained from the CPD of T as follows:*

$$\psi(T) = USV^{\top},$$

where

$$U = V^{(2k-1)} \odot V^{(2k-3)} \odot \cdots \odot V^{(1)}, \qquad V = V^{(2k)} \odot V^{(2k-2)} \odot \cdots \odot V^{(2)},$$

and $S \in \mathbb{R}^{r \times r}$ is a diagonal matrix containing the weights of the CPD on its diagonal (r is the CP rank of T).

The proof of the lemma can be found in [3].

Corollary 3.2 ([3]) *Suppose that A satisfies the condition stated in Lemma 3.1. The equilibrium point $X_e = 0$ of the EPDS (3.3) is asymptotically stable if the first weight element of the CPD is less than one.*

Proof From linear algebra, it is known that the spectral radius of a matrix is always less than or equal to its largest singular value. Therefore, by Lemma 3.1 and Proposition 3.1, the result follows immediately. $\qquad\qquad\square$

Corollary 3.3 ([3]) *Suppose that $\bar{A}$ is given in the TTD form with first $k - 1$ core tensors left-orthonormal and the last k core tensors right-orthonormal, where $\bar{A}$ is the $\mathbb{S}$-transpose of A with $\mathbb{S} = \{1, 3, \ldots, 2k - 1, 2, 4, \ldots, 2k\}$. The equilibrium point $X_e = 0$ of the EPDS (3.3) is asymptotically stable if the largest singular value of $V^{(k)}_{(2,\{1,2,3\})}$, the left-unfolded matrix of the factor tensor $V^{(k)}$ of $\bar{A}$, is less than one.*

Proof The U-singular values of A are the singular values of $\mathbf{V}^{(k)}_{(2,1,2,3)}$ according to Algorithm 1. Therefore, the result follows immediately from Proposition 3.1. $\square$

Remark 3.1 Although Proposition 3.1 provides strong stability results, the three corollaries can efficiently determine the asymptotic stability if A is given in the HOSVD, CPD, or TTD form. For instance, finding the TTD of $\bar{\mathsf{A}}$, left-/right-orthonormalization, and computing the largest singular value of $\mathbf{V}^{(k)}_{(2,1,2,3)}$ can be done in a total time complexity of about $\mathcal{O}(knr^3)$, assuming $n_p = n$ and the TT-ranks $r_p = r$ [3, 8].

3.2.2 Reachability

The framework for analyzing the reachability of the EPDS (3.1, top) can be formulated similarly to that of linear systems theory.

Proposition 3.2 ([3, 4]) *The EPDS (3.1, top) is reachable if and only if the reachability Gramian defined as*

$$W_R = \sum_{t=t_0}^{t_f-1} \mathsf{A}^{t_f-t-1} * \mathsf{B} * \mathsf{B}^\top * (\mathsf{A}^\top)^{t_f-t-1} \in \mathbb{R}^{n_1 \times n_1 \times \cdots \times n_k \times n_k} \tag{3.4}$$

is U-positive definite (i.e., $\psi(W_R)$ is positive definite).

Proof The result can be obtained by utilizing the properties of tensor unfolding ψ and applying concepts from linear systems theory. $\square$

Note that A^{t_f-t-1} denotes the $(t_f - t - 1)$th Einstein power of A. Analogous to LDSs, the reachability Gramian of an EPDS can be computed through a tensor-based Lyapunov equation, which is defined as

$$W_R - \mathsf{A} * W_R * \mathsf{A}^\top = \mathsf{B} * \mathsf{B}^\top. \tag{3.5}$$

Moreover, a tensor-based Kalman's rank condition can be derived for EPDSs.

Proposition 3.3 ([3, 4]) *The EPDS (3.1, top) is reachable if and only if the reachability tensor (a generalized row block tensor) defined as*

$$R = \left[\mathsf{B}\ \mathsf{A} * \mathsf{B} \cdots \mathsf{A}^{\prod_{p=1}^{k} n_p - 1} * \mathsf{B} \right] \in \mathbb{R}^{n_1 \times n_1 s_1 \times \cdots \times n_k \times n_k s_k} \tag{3.6}$$

spans $\mathbb{R}^{n_1 \times n_2 \times \cdots \times n_k}$. In other words, $\mathrm{rank}(\psi(R)) = \prod_{p=1}^{k} n_p$.

The proof can be readily formulated using the tensor unfolding ψ. When $k = 1$, Proposition 3.3 reduces to the famous Kalman's rank condition for LDSs. The following discussion utilizes tensor decompositions, including HOSVD, CPD, and TTD, to establish necessary or sufficient conditions for determining the reachability of EPDSs.

Lemma 3.2 ([3]) *Suppose that $T \in \mathbb{R}^{n_1 \times m_1 \times \cdots \times n_k \times m_k}$. If the rank of $\psi(T)$ is equal to $\prod_{p=1}^{k} n_p$, then the $(2p-1)$-rank of A is equal to n_p for all p. If the rank of $\psi(T)$ is equal to $\prod_{p=1}^{k} m_p$, then the $2p$-rank of T is equal to m_p for all p.*

The proof of the lemma can be found in [3].

Corollary 3.4 ([3]) *Suppose that R is given in the HOSVD form. If the $(2p-1)$-rank of R is not equal to n_p for one p, then the EPDS (3.1) is not reachable. In other words, if the set of $(2p-1)$-mode singular values of R includes zero for one p, then the EPDS (3.1) is not reachable.*

Proof The first part of the result follows immediately from Proposition 3.3 and Lemma 3.2. The second part of the result is based on the fact that the p-rank of a tensor is equal to the number of nonzero p-mode singular values [9]. $\square$

Lemma 3.3 ([3]) *Suppose that $T \in \mathbb{R}^{n_1 \times m_1 \times \cdots \times n_k \times m_k}$ is given in the CPD form with CP rank r. If its factor matrices $V^{(p)}$ and $V^{(q)}$ satisfy*

$$\sum_{\text{odd } p} Krank(V^{(p)}) \geq r + k - 1, \qquad \sum_{\text{even } p} Krank(V^{(q)}) \geq r + k - 1, \tag{3.7}$$

and the Krank of all the factor matrices must be greater than or equal to one, then $rank\big(\psi(T)\big) = r$.

The proof of the lemma can be found in [3].

Corollary 3.5 ([3]) *If R is given in the CPD form with CP rank $\prod_{p=1}^{k} n_p$ and the factor matrices satisfying (3.7), then the EPDS (3.1) is reachable. Conversely, if the EPDS (3.1) is reachable, then the CP rank of R is greater than or equal to $\prod_{p=1}^{k} n_p$.*

Proof The first part of the result can be obtained directly from Proposition 3.3 and Lemma 3.3. The second part of the result relies on the fact that the CP rank of a tensor is greater than or equal to the rank of any unfolded matrix of the tensor [10]. $\square$

Corollary 3.6 ([3]) *Suppose that $\bar{R}$ is given in the TTD form where $\bar{R}$ is the $\mathbb{S}$-transpose of R with $\mathbb{S} = \{1, 3, \ldots, 2k - 1, 2, 4, \ldots, 2k\}$. The EPDS (3.1) is reachable if and only if the kth optimal TT-rank of $\bar{R}$ is equal to $\prod_{p=1}^{k} n_p$.*

Proof The result can be derived using Proposition 3.3 and the definition of optimal TT-ranks. $\qquad\square$

Remark 3.2 Computing the rank of $\psi(R)$ using QR decomposition can be computationally expensive, with a time complexity of $\mathcal{O}(\prod_{p=1}^{k} n_p^3 s_p)$. However, if R is given in the HOSVD, CPD, or TTD form, determining the reachability of EPDSs can be achieved efficiently using different notions of tensor rank. Notably, determining the rank of $\psi(R)$ only requires finding the TTD of $\bar{R}$ [3].

3.2.3 Observability

The observability framework for the EPDS (3.1) can be obtained from the duality principle, which arises from the properties of the unfolding ψ.

Proposition 3.4 ([3, 4]) *The EPDS (3.1) is observable if and only if the observability Gramian defined as*

$$W_O = \sum_{t=t_0}^{t_f - 1} (A^\top)^{t - t_0} * C^\top * C * A^{t - t_0} \in \mathbb{R}^{n_1 \times n_1 \times \cdots \times n_k \times n_k} \tag{3.8}$$

is U-positive definite (i.e., $\psi(W_O)$ is positive definite).

Similarly, the observability Gramian can be computed from a tensor-based Lyapunov equation, i.e.,

$$A^\top * W_O * A - W_O = -C^\top * C. \tag{3.9}$$

Proposition 3.5 ([3, 4]) *The EPDS (3.1) is observable if and only if the observability tensor (a generalized column block tensor) defined as*

$$O = \left[C \ \ C * A \ \cdots \ B * A^{\prod_{p=1}^{k} n_p - 1} \right]^\top \in \mathbb{R}^{n_1 m_1 \times n_1 \times \cdots \times n_k m_k \times n_k} \tag{3.10}$$

spans $\mathbb{R}^{n_1 \times n_2 \times \cdots \times n_k}$. In other words, $\operatorname{rank}\bigl(\psi(O)\bigr) = \prod_{p=1}^{k} n_p$.

Corollary 3.7 ([3]) *Suppose that O is given in the HOSVD form. If the $2p$-rank of O is not equal to n_p for one p, then the EPDS (3.1) is not observable. In other words, if the* *set of $2p$-mode singular values of O contains zero for one p, then the EPDS (3.1) is not observable.*

Corollary 3.8 ([3]) *If O is given in the CPD form with CP rank $\prod_{p=1}^{k} n_p$ and factor matrices satisfying (3.7), then the EPDS (3.1) is observable. Conversely, if the EPDS (3.1) is observable, then the CP rank of O is greater than or equal to $\prod_{p=1}^{k} n_p$.*

Corollary 3.9 ([3]) *Suppose that $\bar{O}$ is given in the TTD form where $\bar{O}$ is the $\mathbb{S}$-transpose of O with $\mathbb{S} = \{1, 3, \dots, 2k - 1, 2, 4, \dots, 2k\}$. The EPDS (3.1) is observable if and only if the kth optimal TT-rank of $\bar{O}$ is equal to $\prod_{p=1}^{k} n_p$.*

3.3   Model Reduction/System Identification

In order to apply standard model reduction/identification frameworks, such as balanced truncation [11, 12], balanced proper orthogonal decomposition [13, 14], and eigensystem realization algorithm [15, 16], to EPDSs, one must use their unfolded representation (3.2), which can pose significant computational challenges. This section presents a tensor-based computational framework for model reduction and system identification of EPDSs [8]. In particular, TTD is utilized as the underlying computational framework. For convenience, the final reduced model is expressed in the unfolded representation, i.e., the LDS model.

3.3.1   TTD-Based Balanced Truncation

Balanced truncation (BT) is a widely used model reduction method for stable linear input-output systems, first proposed by Moore [11] in the 1980s. It has the ability to preserve important system-theoretic properties such as stability and passivity [17]. However, the computational and memory complexity of the standard BT can be prohibitive, with time complexity of $\mathcal{O}(n^3)$ and memory complexity of $\mathcal{O}(n^2)$, where n is the dimension of the state variable [12]. To mitigate these challenges, TTD can be utilized for efficient computation and low storage consumption in BT. This approach is referred to as TTD-based BT.

The TTD-based BT approach also involves solving the tensor-based Lyapunov equations (3.5) and (3.9) as the first step. For solving these equations in BT, several iterative methods have been proposed, such as the Smith method [18–20] and alternating direction implicit methods [21–23], which are suitable for large-scale problems. The tensor-based algebraic Lyapunov equation can be expressed in a normal form as:

$$\mathsf{X} - \mathsf{A} * \mathsf{X} * \mathsf{A}^{\top} = \mathsf{D}. \tag{3.11}$$

The normal form can be reformulated as the following multilinear system:

$$(I - A \circ A) * \bar{X} = \bar{D}, \tag{3.12}$$

where I is the U-identity tensor (i.e., $\psi(I)$ is the identity matrix), and $\bar{X}$ and $\bar{D}$ are the S-transpose of X and D, respectively, with $\mathbb{S} = \{1, 3, \ldots, 2k - 1, 2, 4, \ldots, 2k\}$. If A and D are provided in the TTD form with low TT-ranks, the multilinear system (3.12) can be solved efficiently by exploiting Density Matrix Renormalization Group (DMRG)-based algorithms [24–26]. A similar approach for solving the standard Lyapunov equations was used in [27]. Unfortunately, there is no rigorous theoretical convergence analysis for DMRG-based algorithms [26].

After obtaining the reachability and observability Gramians $\bar{W}_R$ and $\bar{W}_O$ in the TTD form, the economy-size TSVD (using Algorithm 1) can be applied to find the Cholesky-like factors of the two Gramians, i.e.,

$$W_R = Z_R * Z_R^\top \text{ and } W_O = Z_O * Z_O^\top,$$

where $Z_R \in \mathbb{R}^{n_1 \times r_1 \times \cdots \times n_k \times r_k}$ and $Z_O \in \mathbb{R}^{n_1 \times o_1 \times \cdots \times n_k \times o_k}$ are the generalized row block tensors that contain U-singular tensors multiplied by the square root of the corresponding U-singular values. Since the numbers of non-vanishing U-singular values $\prod_{p=1}^{k} r_p$ and $\prod_{p=1}^{k} o_p$ are typically small after the economy-size TSVD, it is more convenient to conduct the rest of the computation in the matrix space. Therefore, the Hankel matrix is computed as

$$\mathbf{H} = \psi(Z_O)^\top \psi(Z_R) \in \mathbb{R}^{\prod_{p=1}^{k} o_p \times \prod_{p=1}^{k} r_p}. \tag{3.13}$$

The remaining steps of the TTD-based BT proceed similarly to the standard BT. The detailed procedure of the TTD-based BT is summarized in Algorithm 2. It should be noted that quantized TTD can also be used in Step 2 to further accelerate the computation of the TTD-based BT.

The standard BT provides an a priori error bound for the reduced system based on the Hankel singular values [11]. However, this error bound does not hold exactly for the TTD-based BT, as several truncation errors occur in the computations of generalized TTD, multilinear system, and economy-size TSVD. Let G and $\mathbf{G}_{\text{red}}$ denote the transfer functions of the original EPDS and the reduced unfolding system, respectively (where G can be defined via the unfolding ψ). The error bound for the TTD-based BT can be written as

$$\|\psi(G) - \mathbf{G}_{\text{red}}\|_\infty \leq \sum_{j=h+1}^{\prod_{p=1}^{k} n_p} \sigma_j + \epsilon, \tag{3.14}$$

where σ_j are the Hankel singular values, h is the matrix SVD truncation point in Step 6, and ϵ is an error term resulting from the TTD-related computations. Here $\| \cdot \|_\infty$ denotes the $\mathcal{H}$-∞ norm. As long as A, B, and C have low TT-ranks, the error term ϵ can be considered negligible, and Algorithm 2 is more computationally efficient than the standard BT.

Algorithm 2 TTD-based Balanced Truncation. This algorithm was adapted from [8] with permission.

1: Given the EPDS (3.1) with $\mathsf{A} \in \mathbb{R}^{n_1 \times n_1 \times \cdots \times n_k \times n_k}$, $\mathsf{B} \in \mathbb{R}^{n_1 \times s_1 \times \cdots \times n_k \times s_k}$, and $\mathsf{C} \in \mathbb{R}^{m_1 \times n_1 \times \cdots \times m_k \times n_k}$

2: Compute the generalized TTDs of A, B, and C

3: Compute the reachability and observability Gramians $\bar{\mathsf{W}}_R \in \mathbb{R}^{n_1 \times n_2 \times \cdots \times n_k \times n_1 \times n_2 \times \cdots \times n_k}$ and $\bar{\mathsf{W}}_O \in \mathbb{R}^{n_1 \times n_2 \times \cdots \times n_k \times n_1 \times n_2 \times \cdots \times n_k}$ in the TTD form using (3.12) via DMRG-based algorithms.

4: Apply Step 3-7 from Algorithm 1 to obtain Z_R and Z_O in the TTD form

5: Recover Z_R and Z_O into the full representation, and compute the Hankel matrix $\mathbf{H}$ using (3.13)

6: Compute the economy-size matrix SVD of $\mathbf{H}$, i.e., $\mathbf{H} \approx \mathbf{U}\mathbf{S}\mathbf{V}^\top$ where $\mathbf{U} \in \mathbb{R}^{\prod_{p=1}^{k} o_p \times h}$, $\mathbf{S} \in \mathbb{R}^{h \times h}$, and $\mathbf{V} \in \mathbb{R}^{\prod_{p=1}^{k} r_p \times h}$ (h is the number of the non-vanishing Hankel singular values)

7: Calculate the transformation matrices

$$\mathbf{P} = \mathbf{Z}_R \mathbf{V}\mathbf{S}^{-\frac{1}{2}} \in \mathbb{R}^{\prod_{p=1}^{k} n_p \times h} \quad \text{and} \quad \mathbf{Q} = \mathbf{Z}_O \mathbf{U}\mathbf{S}^{-\frac{1}{2}} \in \mathbb{R}^{\prod_{p=1}^{k} n_p \times h}$$

8: The reduced unfolded model is computed as

$$\mathbf{A} = \mathbf{Q}^\top \psi(\mathsf{A})\mathbf{P} \in \mathbb{R}^{h \times h}$$

$$\mathbf{B} = \mathbf{Q}^\top \psi(\mathsf{B}) \in \mathbb{R}^{h \times \prod_{p=1}^{k} s_p}$$

$$\mathbf{C} = \psi(\mathsf{C})\mathbf{P} \in \mathbb{R}^{\prod_{p=1}^{k} m_p \times h}$$

9: **return** The reduced model of the EPDS.

3.3.2   TTD-Based Balanced Proper Orthogonal Decomposition

The reachability and observability Gramians can be directly obtained from data or numerical simulations, bypassing the need to solve the tensor Lyapunov equations, as shown in [13]. Lall et al. [28, 29] originally used this approach to generalize BT to nonlinear systems. Given the EPDS (3.1, top), the even-order control tensor B can be decomposed as

$$\mathsf{B} = \begin{bmatrix} \mathsf{B}_1 & \mathsf{B}_2 & \cdots & \mathsf{B}_{\prod_{p=1}^{k} s_p} \end{bmatrix}$$

using the inverse operation of the generalized row block tensor construction. Then the state responses can be constructed for $t = 1, 2, \ldots, r$ as follows:

$$\mathsf{X}^{(j)} = \begin{bmatrix} \mathsf{B}_j & \mathsf{A} * \mathsf{B}_j & \cdots & \mathsf{A}^r * \mathsf{B}_j \end{bmatrix}.$$

If A and B are already given in the generalized TTD form with low TT-ranks, then the state responses $\mathsf{X}^{(j)}$ can be efficiently constructed in the generalized TTD form. Consider $r + 1 = \prod_{p=1}^{k} r_p$ snapshots of states for $j = 1, 2, \ldots, \prod_{p=1}^{k} s_p$ and arrange the snapshots in the following form:

$$\mathsf{X} = \left[\mathsf{X}^{(1)}\ \mathsf{X}^{(2)}\ \ldots\ \mathsf{X}^{(\prod_{p=1}^{k} s_p)}\right] \in \mathbb{R}^{n_1 \times s_1 r_1 \times \cdots \times n_k \times s_k r_k}. \tag{3.15}$$

The empirical reachability Gramian of the EPDS (3.1) can be computed as

$$\mathsf{W}_R = \mathsf{X} * \mathsf{X}^\top. \tag{3.16}$$

To construct the empirical observability Gramian $\mathsf{W}_O = \mathsf{Y}^\top * \mathsf{Y}$, the adjoint system of the EPDS (3.1) is solved, following a procedure similar to that for the reachability Gramian, i.e.,

$$\tilde{\mathsf{X}}_{t+1} = \mathsf{A}^\top * \tilde{\mathsf{X}}_t + \mathsf{C}^\top * \mathsf{V}_t$$

over $t = 1, 2, \ldots, o$ to obtain the empirical observability snapshot tensor Y (in the generalized TTD form). After obtaining both the empirical reachability and observability snapshot tensors X and Y, the generalized Hankel tensor can be constructed as

$$\mathsf{H} = \mathsf{Y}^\top * \mathsf{X} \in \mathbb{R}^{m_1 o_1 \times s_1 r_1 \times \cdots \times m_k o_k \times s_k r_k} \tag{3.17}$$

in the generalized TTD form, where $o + 1 = \prod_{p=1}^{k} o_p$. Finally, Algorithm 1 is applied to obtain the generalized Hankel singular values, and proceed with the remaining steps of the TTD-based balanced proper orthogonal decomposition (BPOD) as for the TTD-based BT. The detailed steps of the TTD-based BPOD are summarized in Algorithm 3. Note that quantized TTD can also be used in Step 2 to further accelerate the computation of the TTD-based BPOD.

Algorithm 3 TTD-based Balanced Proper Orthogonal Decomposition. This algorithm was adapted from [8] with permission.

1: Given the EPDS (3.1) with $\mathsf{A} \in \mathbb{R}^{n_1 \times n_1 \times \cdots \times n_k \times n_k}$, $\mathsf{B} \in \mathbb{R}^{n_1 \times s_1 \times \cdots \times n_k \times s_k}$, and $\mathsf{C} \in \mathbb{R}^{m_1 \times n_1 \times \cdots \times m_k \times n_k}$, and two integer r and o

2: Find the generalized TTDs of A, B, and C

3: Construct the reachability and observability snapshot tensors X and Y using (3.15) in the generalized TTD form

4: Compute the generalized Hankel tensor H using (3.17) in the TTD form

5: Compute the economy-size TSVD of H using Algorithm 1, i.e., $\mathsf{H} = \mathsf{U} * \mathsf{S} * \mathsf{V}^\top$ and unfold the TSVD to the matrix form using ψ, i.e., $\mathbf{H} \approx \mathbf{U}\mathbf{S}\mathbf{V}^\top$ where $\mathbf{U} \in \mathbb{R}^{\prod_{p=1}^{k} m_p o_p \times h}$, $\mathbf{S} \in \mathbb{R}^{h \times h}$, and $\mathbf{V} \in \mathbb{R}^{\prod_{p=1}^{k} s_p r_p \times h}$

6: Follow similarly as Steps 7 and 8 in Algorithm 2

7: **return** The reduced model of the EPDS.

The TTD-based BPOD is not suitable for experimental settings since it requires access to the EPDS and its adjoint. Additionally, constructing the generalized Hankel tensor through the TTD-based Einstein product between the reachability and observability snapshots X and Y can be computationally expensive if the TT-ranks are not adequately small.

3.3.3 TTD-Based Eigensystem Realization Algorithm

Eigensystem realization algorithm (ERA) was initially introduced as a model identification and reduction technique for LDSs in [15]. It has been subsequently proven that ERA is equivalent to BPOD for discrete-time systems, but with much lower computational overhead [30]. Like TTD-based BPOD, TTD-based ERA also seeks to create the generalized Hankel tensor from response simulations.

First, snapshots of the impulse responses, also known as Markov parameters, need to be collected, i.e.,

$$\mathcal{Z}_t = \mathcal{C} * \mathcal{A}^t * \mathcal{B} \in \mathbb{R}^{m_1 \times s_1 \times \cdots \times m_k \times s_k}. \tag{3.18}$$

Next, the generalized Hankel tensor can be constructed using a generalized row/column block tensor approach, i.e.,

---

Algorithm 4 TTD-based Eigensystem Realization Algorithm. This algorithm was adapted from [8] with permission.

1: Given the EPDS (3.1) with $\mathcal{A} \in \mathbb{R}^{n_1 \times n_1 \times \cdots \times n_k \times n_k}$, $\mathcal{B} \in \mathbb{R}^{n_1 \times s_1 \times \cdots \times n_k \times s_k}$, and $\mathcal{C} \in \mathbb{R}^{m_1 \times n_1 \times \cdots \times m_k \times n_k}$, and two integer r and o

2: Find the generalized TTDs of $\mathcal{A}$, $\mathcal{B}$, and $\mathcal{C}$

3: Construct the snapshots of the impulse responses $\mathcal{Z}_t \in \mathbb{R}^{m_1 \times s_1 \times \cdots \times m_k \times s_k}$ in the generalized TTD form

4: Compute the generalized Hankel tensor $\mathcal{H}$ using (3.19) in the TTD form

5: Compute the economy-size TSVD of $\mathcal{H}$ using Algorithm 1, i.e., $\mathcal{H} = \mathcal{U} * \mathcal{S} * \mathcal{V}^\top$ and unfold the TSVD to the matrix form using ψ, i.e., $\mathbf{H} \approx \mathbf{U}\mathbf{S}\mathbf{V}^\top$ where $\mathbf{U} \in \mathbb{R}^{\prod_{p=1}^{k} m_p o_p \times h}$, $\mathbf{S} \in \mathbb{R}^{h \times h}$, and $\mathbf{V} \in \mathbb{R}^{\prod_{p=1}^{k} s_p r_p \times h}$

6: The reduced model is given by

$$\mathbf{A} = \mathbf{S}^{-\frac{1}{2}} \mathbf{U}^\top \mathbf{H}_1 \mathbf{V}\mathbf{S}^{-\frac{1}{2}} \in \mathbb{R}^{h \times h},$$

$$\mathbf{B} = \mathbf{S}^{-\frac{1}{2}} \mathbf{U}^\top \mathbf{H}_{\mathrm{col}} \in \mathbb{R}^{h \times \prod_{p=1}^{k} s_p},$$

$$\mathbf{C} = \mathbf{H}_{\mathrm{row}} \mathbf{V}\mathbf{S}^{-\frac{1}{2}} \in \mathbb{R}^{\prod_{p=1}^{k} m_p \times h},$$

where

$$\mathbf{H}_{\mathrm{col}} = \psi\left(\begin{bmatrix} \mathcal{Z}_0 & \mathcal{Z}_1 & \cdots & \mathcal{Z}_S \end{bmatrix}^\top \right), \quad \mathbf{H}_{\mathrm{row}} = \psi\left(\begin{bmatrix} \mathcal{Z}_0 & \mathcal{Z}_1 & \cdots & \mathcal{Z}_T \end{bmatrix} \right),$$

$$\mathbf{H}_1 = \psi\left(\begin{bmatrix} \mathcal{Z}_1 & \mathcal{Z}_2 & \cdots & \mathcal{Z}_{T+1} \\ \mathcal{Z}_2 & \mathcal{Z}_3 & \cdots & \mathcal{Z}_{T+2} \\ \vdots & \vdots & \cdots & \vdots \\ \mathcal{Z}_{S+1} & \mathcal{Z}_{S+2} & \cdots & \mathcal{Z}_{T+S+1} \end{bmatrix} \right)$$

7: **return** The reduced model of the EPDS.

$$H = \begin{bmatrix} Z_0 & Z_1 & \cdots & Z_r \\ Z_1 & Z_2 & \cdots & Z_{r+1} \\ \vdots & \vdots & \ddots & \vdots \\ Z_o & Z_{o+1} & \cdots & Z_{r+o} \end{bmatrix} \in \mathbb{R}^{m_1 o_1 \times s_1 r_1 \times \cdots \times m_k o_k \times s_k r_k}, \tag{3.19}$$

where $r + 1 = \prod p = 1^k r_p$ and $o + 1 = \prod p = 1^k o_p$. The detailed procedure of the TTD-based ERA can be found in Algorithm 4. Again, quantized TTD can also be used in Step 2 to further accelerate the computation of the TTD-based ERA. More significantly, the Markov parameters Z_t can be obtained from impulse responses without having access to the EPDS. Therefore, if the Markov parameters Z_t are obtained from experiments, the TTD-based ERA can begin at Step 4.

3.4 Applications

3.4.1 Synthetic Data: Reachability and Observability

This example was adapted from [3]. Consider the same example as presented in Chap. 2. The TPDS can be equivalently represented in the form (3.1) with $A = A_1 \circ A_2, B = B_1 \circ B_2$, and $C = C_1 \circ C_2$. Based on Propositions 3.3 and 3.5, the reachability and observability tensors are computed as

$$R_{::11} = \begin{bmatrix} 0 & 0 & 0 \\ 0 & 1 & 0 \\ 0 & 0.8 & 0 \end{bmatrix}, \qquad R_{::12} = \begin{bmatrix} 0.4 & 0 & 0.378 \\ 0.57 & 0 & 0.4849 \\ 0.756 & 0 & 0.6339 \end{bmatrix},$$

$$R_{::21} = \begin{bmatrix} 0 & 0 & 0.5 \\ 0 & 0 & 0.4 \\ 1 & 0 & 0.57 \end{bmatrix}, \qquad R_{::22} = \begin{bmatrix} 0 & 0.285 & 0 \\ 0 & 0.378 & 0 \\ 0 & 0.4849 & 0 \end{bmatrix},$$

$$O_{::11} = \begin{bmatrix} 1 & 0 & 0 \\ 0 & 0 & 0 \\ 0 & 0 & 0.5 \end{bmatrix}, \qquad O_{::12} = \begin{bmatrix} 0 & 0 & 0 \\ 0 & 1 & 0 \\ 0 & 0 & 0 \end{bmatrix},$$

$$O_{::21} = \begin{bmatrix} 0 & 0 & 0 \\ 0.04 & 0.15 & 0.285 \\ 0 & 0 & 0 \end{bmatrix}, \quad O_{::22} = \begin{bmatrix} 0.1 & 0.25 & 0.4 \\ 0 & 0 & 0 \\ 0.057 & 0.1825 & 0.378 \end{bmatrix},$$

respectively. It is clear that $\psi(R) = \mathbf{R}$ and $\psi(O) = \mathbf{O}$, where $\mathbf{R}$ and $\mathbf{O}$ are the reachability and observability matrices defined in the previous example. The second optimal TT-ranks of $\bar{R}$ and $\bar{O}$ are equal to 6, as defined in Corollary 3.3. Hence, the system is both reachable and observable.

3.4.2   Synthetic Data: Stability

This example, adapted from [3], aims to investigate the computational efficiency of the TTD-based TSVD (Corollary 3.3) with the unfolding-based TSVD (with the tensor unfolding ψ) when determining the stability of EPDSs. Consider unforced EPDSs with random sparse even-order ($2k$th-order) dynamic tensors $\mathsf{A} \in \mathbb{R}^{(2\times2)\times\overset{k}{\cdots}\times(2\times2)}$ with different orders such that the TTD of A is provided. The results are shown in Table 3.1, indicating that the TTD-based TSVD is significantly faster than the matrix SVD-based TSVD in finding the largest U-singular value of A for $k \geq 10$. The computational time for the matrix SVD-based method increases exponentially as k increases.

3.4.3   Synthetic Data: TTD-Based Balanced Truncation

This example, adapted from [8], aims to compare the computational time of the TTD-based BT (Algorithm 2) with the standard BT. Consider multiple-input and multiple-output EPDSs with random sparse even-order ($2k$th-order) parameter tensors $\mathsf{A} \in \mathbb{R}^{(2\times2)\times\overset{k}{\cdots}\times(2\times2)}$, $\mathsf{B} \in \mathbb{R}^{(2\times2)\times\overset{k}{\cdots}\times(2\times2)}$, and $\mathsf{C} \in \mathbb{R}^{(2\times2)\times\overset{k}{\cdots}\times(2\times2)}$ with different orders such that the generalized TTDs of A, B, and C are provided. The computational time for the standard BT grows rapidly with the number of states, i.e., 2^k, as shown in Fig. 3.1. In contrast, the TTD-based BT has a bounded computational time. Furthermore, the error bounds in model reduction using both methods are very close to each other. Note that for the TTD-based BT, the error bounds are computed based on the sum of the residual Hankel singular values and the $\mathcal{H}\text{-}\infty$ norm between the two transfer functions from the standard and the TTD-based reduced models. This norm estimates the error term ϵ in (3.14) based on the triangle inequality.

Table 3.1 Computational time (in second) comparison between the TTD-based and matrix SVD-based TSVD in finding the largest U-singular value of A. This table was adapted from [3] with permission

Order k	TTD-based	Unfolding-based	$\sigma_{\max}$	Relative error	Stability
6	0.0399	6.8551×10^{-4}	0.8082	1.3738×10^{-16}	Asymptotically stable
8	0.0491	0.0439	0.9626	4.1523×10^{-15}	Asymptotically stable
10	0.0591	0.4979	0.8645	3.8527×10^{-15}	Asymptotically stable
12	0.0909	30.7663	0.8485	5.7573×10^{-15}	Asymptotically stable
14	0.2623	2115.1	0.9984	1.3566×10^{-14}	Asymptotically stable

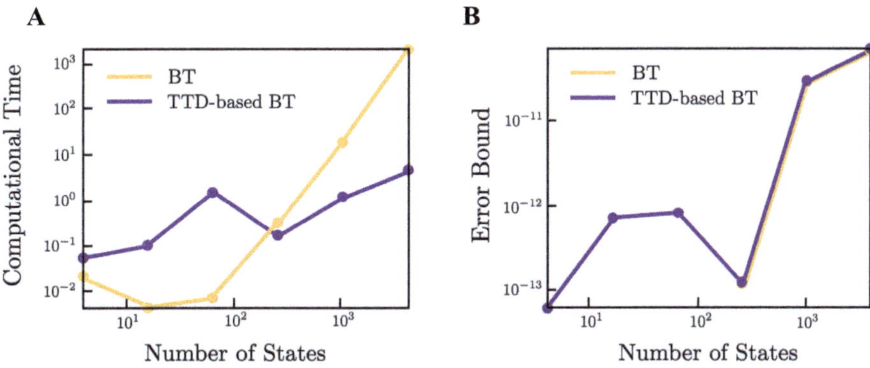

Fig. 3.1 Computational time in second (**A**) and error bound (**B**) comparisons between the standard and the TTD-based BT. The error bounds for the first three states of the standard BT are zeros. This figure was redrawn from [8] with permission

3.4.4 2D Heat Equations with Control

This example was adapted from [8]. The heat equation is a partial differential equation that models the diffusion of heat in a solid medium over time. [31, 32]. For a domain $D = [-\pi, \pi] \times [-\pi, \pi]$ with localized point control and Dirichlet boundary conditions, the two-dimensional heat equation is defined as

$$\begin{cases} \frac{\partial}{\partial t}\phi(t, \mathbf{x}) = c^2 \frac{\partial^2}{\partial \mathbf{x}^2}\phi(t, \mathbf{x}) + \delta(\mathbf{x})u_t, \ \mathbf{x} \in D \\ \phi(\mathbf{x}, t) = 0, \ \mathbf{x} \in \partial D \end{cases},$$

where $c > 0$, $u_t \in \mathbb{R}$ is the control input, and $\delta(\mathbf{x})$ is the Dirac delta function centered at zero.

This example compares the computational time of the TTD-based BPOD (Algorithm 3) and the TTD-based ERA (Algorithm 4) with their corresponding standard methods on the EPDSs obtained from the discretized heat equation. To discretize the heat equation, a second-order central difference scheme is used to estimate the Laplacian and approximated the Dirac delta as a Kronecker delta function at the nearest grid point with unit mass. This results in EPDSs with $\mathsf{A} \in \mathbb{R}^{n \times n \times n \times n}$, $\mathbf{B} \in \mathbb{R}^{n \times n}$, and $\mathbf{C} \in \mathbb{R}^{n \times n}$, where $n = \frac{2\pi}{h}$ (assuming measurements are taken at a single discrete location and the spatial resolutions of grid sizes are equal to h for both dimensions). To ensure numerical stability, the Courant-Friedrichs-Lewy condition $\frac{c^2 \Delta t}{h^2} < 1$ (Δt is the discretization in time) is met [33].

Both generalized TTD and quantized TTD (including TTD conversion time) are compared for the TTD-based BPOD/ERA. The results are shown in Fig. 3.2A and C, where both TTD-based methods are found to be more efficient than their corresponding standard methods as the number of states, i.e., n^2, becomes large. In particular, using the quantized TTD is even faster than using the generalized TTD when the conversion time is included. However,

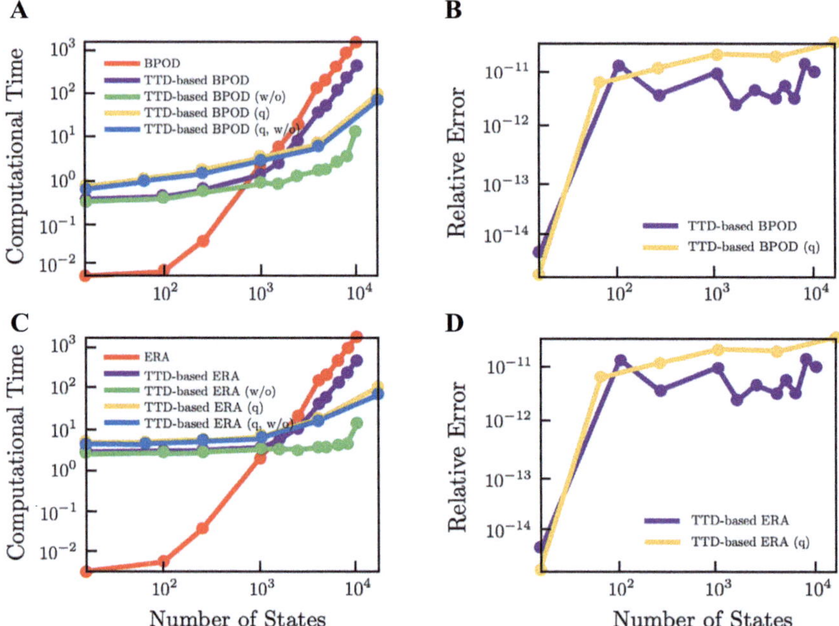

Fig. 3.2 Computational time in second (**A**) and (**C**) and relative error (**B**) and (**D**) comparisons between the TTD-based BPOD/ERA with their corresponding standard methods. "q" stands for the case using quantized TTD, and "w/o" stands for the case where the conversion time to generalized/quantized TTD is omitted. This figure was drawn from [8] with permission

for special tensors like the Laplacian that can be directly and efficiently constructed in the generalized/quantized TTD form, the generalized TTD outperforms the quantized TTD, as shown in the curves where the conversion time is not included. Figure 3.2B and D show the relative error based on the $\mathcal{H}$-∞ norm between the transfer functions obtained from the two TTD-based methods and their corresponding standard methods. Clearly, the TTD-based methods can obtain reduced models with similar accuracy (compared to the standard methods) and lower computational costs.

3.4.5   Room Impulsive Response Dataset

This example was adapted from [8]. Reducing room reverberation is of great importance in improving the quality of human life. This example demonstrates the application of the TTD-based ERA to a real-world database of binaural room impulse responses, known as the Aachen Impulse Response (AIR) database [34, 35]. The impulse response in this case is one-dimensional, so the standard ERA can be applied directly. However, the generalized Hankel matrices built from the impulse responses are very large and contain very small

Table 3.2 Computational time comparison (in second) between the standard and the TTD-based ERA for different values of k with sampling frequency 5600. This table was adapted from [8] with permission

Order k	$k = 9$	$k = 10$	$k = 11$	$k = 12$
Threshold ϵ	0.9	1.2	1.5	2.4
ERA	0.1961	0.9432	7.1086	50.6742
TTD-based ERA	0.5075	2.2262	8.2137	10.5300
ERA relative error	0.0890	0.0715	0.0667	0.0666
TTD-based ERA relative error	0.0929	0.0972	0.0932	0.0849

values, making the TTD-based ERA more promising in estimating the acoustic dynamics, especially with the use of quantized TTD.

The AIR database has many scenarios. The focus is on the case where the room impulse responses are collected by a bottom microphone with a bottom-top mock-up phone in an office. A comparison is made between the computational time of the TTD-based ERA (Algorithm 4 from Step 4 with quantized TTD) and the standard ERA for identifying acoustic dynamics. The conversion time to the quantized TTD is included for the TTD-based ERA.

To tune the threshold ϵ in the TTD conversion from H to $\bar{\mathsf{H}}$, the number of snapshots $r = o = 2^k - 1$ is set to a positive integer k. The results are presented in Tables 3.2 and 3.3. In the first case, k is increased to allow for larger truncation thresholds ϵ, facilitating the computation of the economy-size TSVD. The TTD-based ERA shows a computational advantage in approximating the acoustic dynamics with low relative errors for $k \geq 12$, as shown in Table 3.2. Of course, the standard ERA can achieve lower relative errors with the

Table 3.3 Computational time comparison (in second) between the standard and the TTD-based ERA for different truncation thresholds with $k = 13$ and sampling frequency 12000. "# of Singular Values" represents the minimum number of singular values required to maintain the systems with relative good accuracy for the TTD-based ERA. This table was adapted from [8] with permission

Threshold	$\epsilon = 1$	$\epsilon = 2$	$\epsilon = 5$	$\epsilon = 10$
# of Singular values	511	297	72	41
ERA	336.5543	336.5543	336.5543	336.5543
TTD-based ERA	198.5078	105.8521	40.5420	17.3069
ERA Relative lrror	0.0492	0.0835	0.1252	0.1282
TTD-based ERA relative error	0.0721	0.0887	0.1086	0.1215

same number of singular values retained as in the TTD-based ERA. In the second case, when $k = 13$, larger truncation thresholds ϵ enable faster computation and result in similar relative errors, see Table 3.3.

References

1. Mark, Rogers, Lei Li, and Stuart J Russell. 2013. Multilinear dynamical systems for tensor time series. *Advances in Neural Information Processing Systems* 26.
2. Amit, Surana, Geoff Patterson, and Indika Rajapakse. 2016. Dynamic tensor time series modeling and analysis. In *2016 IEEE 55th conference on decision and control (CDC)*, 1637–1642. IEEE.
3. Can, Chen, Amit Surana, Anthony M Bloch, and Indika Rajapakse. 2021. Multilinear control systems theory. *SIAM Journal on Control and Optimization* 59(1):749–776.
4. Can, Chen, Amit Surana, Anthony Bloch, and Indika Rajapakse. 2019. Multilinear time invariant system theory. In *2019 proceedings of the conference on control and its applications*, 118–125. SIAM.
5. Ding, Wenwen, Kai Liu, Evgeny Belyaev, and Fei Cheng. 2018. Tensor-based linear dynamical systems for action recognition from 3d skeletons. *Pattern Recognition* 77: 75–86.
6. Patrick Gelß. 2017. *The tensor-train format and its applications: Modeling and analysis of chemical reaction networks, catalytic processes, fluid flows, and Brownian dynamics*. PhD thesis.
7. Ivan, V. Oseledets. 2010. Approximation of $2^d \times 2^d$ matrices using tensor decomposition. *SIAM Journal on Matrix Analysis and Applications* 31(4):2130–2145.
8. Can, Chen, Amit Surana, Anthony Bloch, and Indika Rajapakse. 2019. Data-driven model reduction for multilinear control systems via tensor trains. arXiv:1912.03569.
9. De Lathauwer, Lieven, Bart De Moor, and Joos Vandewalle. 2000. A multilinear singular value decomposition. *SIAM Journal on Matrix Analysis and Applications* 21 (4): 1253–1278.
10. De Silva, Vin, and Lek-Heng. Lim. 2008. Tensor rank and the ill-posedness of the best low-rank approximation problem. *SIAM Journal on Matrix Analysis and Applications* 30 (3): 1084–1127.
11. Moore, Bruce. 1981. Principal component analysis in linear systems: Controllability, observability, and model reduction. *IEEE Transactions on Automatic Control* 26 (1): 17–32.
12. Serkan, Gugercin and Athanasios C Antoulas. 2004. A survey of model reduction by balanced truncation and some new results. *International Journal of Control* 77(8):748–766.
13. Clarence, W. Rowley. 2005. Model reduction for fluids, using balanced proper orthogonal decomposition. *International Journal of Bifurcation and Chaos* 15(03):997–1013.
14. Miloš, Ilak and Clarence W. Rowley. 2008. Modeling of transitional channel flow using balanced proper orthogonal decomposition. *Physics of Fluids* 20(3):034103.
15. Jer-Nan, Juang, Richard S. Pappa. 1985. An eigensystem realization algorithm for modal parameter identification and model reduction. *Journal of Guidance, Control, and Dynamics* 8(5):620–627.
16. Richard, S. Pappa, Kenny B. Elliott, and Axel Schenk. 1993. Consistent-mode indicator for the eigensystem realization algorithm. *Journal of Guidance, Control, and Dynamics* 16(5):852–858.
17. Pernebo, Lars, and Leonard Silverman. 1982. Model reduction via balanced state space representations. *IEEE Transactions on Automatic Control* 27 (2): 382–387.
18. Smith, R.A. 1968. Matrix equation xa+bx=c. *SIAM Journal on Applied Mathematics* 16 (1): 198–201.
19. Penzl, Thilo. 1999. A cyclic low-rank smith method for large sparse lyapunov equations. *SIAM Journal on Scientific Computing* 21 (4): 1401–1418.

20. Serkan, Gugercin, Danny C. Sorensen, and Athanasios C. Antoulas. 2003. A modified low-rank smith method for large-scale lyapunov equations. *Numerical Algorithms*, 32(1):27–55.
21. Garrett, Birkhoff, Richard S. Varga, and David Young. 1962. Alternating direction implicit methods. In *Advances in computers*, vol. 3, 189–273. Elsevier.
22. Eugene, L. Wachspress. 1988. Iterative solution of the lyapunov matrix equation. *Applied Mathematics Letters* 1(1):87–90.
23. Namiki, Takefumi. 1999. A new FDTD algorithm based on alternating-direction implicit method. *IEEE Transactions on Microwave Theory and Techniques* 47 (10): 2003–2007.
24. Schollwöck, Ulrich. 2005. The density-matrix renormalization group. *Reviews of Modern Physics* 77 (1): 259.
25. Chan, Garnet Kin-Lic, and Sandeep Sharma. 2011. The density matrix renormalization group in quantum chemistry. *Annual Review of Physical Chemistry* 62 (1): 465–481.
26. Ivan, V. Oseledets and Sergey V. Dolgov. 2012. Solution of linear systems and matrix inversion in the tt-format. *SIAM Journal on Scientific Computing* 34(5):A2718–A2739.
27. Michael, Nip, Joao P. Hespanha, and Mustafa Khammash. 2013. Direct numerical solution of algebraic lyapunov equations for large-scale systems using quantized tensor trains. In *52nd IEEE conference on decision and control*, 1950–1957. IEEE.
28. Sanjay, Lall, Jerrold E. Marsden, and Sonja Glavaški. 1999. Empirical model reduction of controlled nonlinear systems. *IFAC Proceedings Volumes* 32(2):2598–2603.
29. Sanjay, Lall, Jerrold E. Marsden, and Sonja Glavaški. 2002. A subspace approach to balanced truncation for model reduction of nonlinear control systems. *International Journal of Robust and Nonlinear Control: IFAC-Affiliated Journal* 12(6):519–535.
30. Zhanhua, Ma, Sunil Ahuja, and Clarence W. Rowley. 2011. Reduced-order models for control of fluids using the eigensystem realization algorithm. *Theoretical and Computational Fluid Dynamics* 25(1):233–247.
31. El-Maati, Ouhabaz. 2009. Analysis of heat equations on domains (lms-31). In *Analysis of Heat Equations on Domains (LMS-31)*. Princeton University Press.
32. Zuazua, Enrique. 2006. Control and numerical approximation of the wave and heat equations. In *International Congress of Mathematicians, Madrid, Spain* 3: 1389–1417.
33. Carlos. A. De Moura and Carlos S. Kubrusly. 2013. The courant–friedrichs–lewy (cfl) condition. *AMC* 10(12).
34. Marco, Jeub, Magnus Schafer, and Peter Vary. 2009. A binaural room impulse response database for the evaluation of dereverberation algorithms. In *2009 16th international conference on digital signal processing*, 1–5. IEEE.
35. Jeub, Marco, Magnus Schäfer, Hauke Krüger, Christoph Nelke, Christophe Beaugeant, and Peter Vary. 2010. Do we need dereverberation for hand-held telephony? In *Proceedings of the international congress on acoustics (ICA)*, Sydney, Australia.

Tensor Vector Product-Based Dynamical Systems 4

Abstract

The tensor vector product-based dynamical system (TVPDS) representation was pioneered in 2021 by Chen et al. [*IEEE Trans. Netw. Sci. Eng.*] as a novel framework for characterizing the multidimensional state dynamics of hypergraphs. Hypergraphs are generalizations of graphs where hyperedges can connect more than two nodes. The evolution of a TVPDS is characterized by the tensor vector product between a dynamic tensor and a state vector, which in fact can be equivalently reformulated in the form of homogeneous polynomial dynamical systems. By utilizing polynomial systems theory and tensor algebra, the system-theoretic properties of TVPDSs, such as stability, controllability, and observability, are investigated. In particular, the controllability and observability of hypergraphs are explored through the lens of the TVPDS representation.

4.1 Overview

Both the TPDS representation (2.1) and the EPDS representation (3.1) are tensor-based dynamical systems that can be transformed into LDSs through tensor unfolding. This transformation allows for the application of well-established linear system analysis techniques to tensor-based dynamical systems. On the other hand, tensor algebra has emerged as a powerful tool for modeling and simulating nonlinear dynamics [1–4]. As mentioned earlier, every homogeneous polynomial can be uniquely represented by a supersymmetric tensor through tensor vector products [5–7]. Therefore, given a kth-order n-dimensional supersymmetric tensor $\mathsf{T} \in \mathbb{R}^{n \times n \times \overset{k}{\cdots} \times n}$, the following product:

$$\mathsf{T} \times_1 \mathbf{v} \times_2 \mathbf{v} \times_3 \cdots \times_{k-1} \mathbf{v} = \mathsf{T}\mathbf{v}^{k-1} \in \mathbb{R}^n$$

belongs to the family of homogeneous polynomial systems. Moreover, if T is only symmetric with respect to the first $k - 1$ modes, referred to as almost symmetric, the above product spans the entire space of homogeneous polynomial systems [2].

Consequently, the continuous-time tensor vector product-based dynamical system (TVPDS) can be defined as

$$\dot{\mathbf{x}}(t) = \mathbf{A}\mathbf{x}(t)^{k-1}, \tag{4.1}$$

which is equivalent to an n-dimensional homogeneous polynomial dynamical system of degree $k - 1$ for almost symmetric dynamic tensor $\mathbf{A} \in \mathbb{R}^{n \times n \times \overset{k}{\cdots} \times n}$. The TVPDS representation serves as a generalization of the LDS model and reduces it as a special case when $k = 2$. Significantly, this representation offers a novel framework to study homogeneous polynomial dynamical systems by leveraging tensor algebra, such as tensor eigenvalues or tensor decompositions. In addition, the TVPDS (4.1) possesses an unfolded form, i.e.,

$$\dot{\mathbf{x}}(t) = \mathbf{A}_{(k)}\big(\mathbf{x}(t) \otimes \mathbf{x}(t) \otimes \overset{k-1}{\cdots} \otimes \mathbf{x}(t)\big), \tag{4.2}$$

where $\mathbf{A}_{(k)} \in \mathbb{R}^{n \times n^{k-1}}$ is the k-mode matricization of $\mathbf{A}$. The discrete-time version of TVPDSs can also be formulated similarly.

TVPDSs can be naturally applied to describe the dynamics of hypergraphs, a generalization of graphs where hyperedges can connect more than two nodes [8], by defining the dynamic tensor $\mathbf{A}$ as the adjacency tensor of the hypergraph [1]. This capability stems from the inherent ability of TVPDSs, namely homogeneous polynomial dynamical systems, to effectively capture higher-order interactions or correlations, making them well-suited for modeling a diverse range of real-world systems [1, 9–13]. For instance, Chen et al. [1] employed TVPDSs to model higher-order interactions in mouse neuronal networks, which offers a unique insight in understanding the functionality of mouse brains. Additionally, Grilli et al. [10] utilized TVPDSs to capture higher-order interactions among communities in complex ecological systems and demonstrated that higher-order interactions have strong effects on the stability of ecological communities. As a matter of fact, every polynomial dynamical system can be homogenized by introducing additional variables, which broadens the applicability of TVPDSs.

However, the nonlinearity inherent in TVPDSs poses challenges in analyzing system-theoretic properties and developing optimal control strategies [2, 14–16]. Local analysis via lineraization remains the most common approach for analyzing TVPDSs [17, 18]. This chapter delves into the recent advancements in system-theoretic properties of TVPDSs, encompassing stability, controllability, and observability, by employing polynomial systems theory and tensor algebra. Furthermore, the concepts of controllability and observability are applied in the context of hypergraph dynamics to identify the minimum number of driver and observable nodes for achieving hypergraph controllability and observability, respectively. The content of this chapter is mainly based on the work of [1, 2, 19, 20].

4.2   System-Theoretic Properties: Stability

The study of stability of TVPDSs, namely homogeneous polynomial dynamical systems, has been an active research area for many decades [2, 14–16]. Recently, Ali and Khadir [16] proved that a homogeneous polynomial dynamical system is asymptotically stable if and only if it admits a rational Lyapunov function. This section discusses the internal stability of a subclass of TVPDSs with odeco dynamic tensors (see Chap. 1 for the definition of odeco tensors). These systems are referred to as odeco TVPDSs.

4.2.1   Explicit Solutions

The explicit solution of an odeco TVPDS is achievable by leveraging the orthogonal decomposition of its dynamic tensor A.

Proposition 4.1 ([2]) *Suppose that the dynamic tensor $A \in \mathbb{R}^{n \times n \times \overset{k}{\cdots} \times n}$ is odeco with the following orthogonal decomposition:*

$$A = \sum_{j=1}^{n} \lambda_j v_j \circ v_j \circ \overset{k}{\cdots} \circ v_j.$$

Let the initial condition $x_0 = \sum_{j=1}^{n} c_j v_j$ (since v_j are orthonormal). Given the initial condition x_0, the explicit solution of the odeco TVDPS (4.1) can be computed as

$$x(t) = \sum_{j=1}^{n} \left(1 - (k-2)\lambda_j c_j^{k-2} t \right)^{-\frac{1}{k-2}} c_j v_j, \tag{4.3}$$

where λ_j are the Z-eigenvalues of A with the corresponding Z-eigenvector v_j.

Proof First, let $\mathbf{x}(t) = \sum_{j=1}^{n} \alpha_j(t) v_j$ with $\alpha_j(0) = c_j$. Based on the property of tensor vector products, it can be shown that

$$\sum_{j=1}^{n} \dot{\alpha}_j(t) \mathbf{v}_j = \left(\sum_{j=1}^{n} \lambda_j v_j \circ v_j \circ \overset{k}{\cdots} \circ v_j \right) \times_1 \left(\sum_{j=1}^{n} \alpha_j(t) v_j \right) \times_2 \cdots \times_{k-1} \left(\sum_{j=1}^{n} \alpha_j(t) v_j \right)$$

$$= \sum_{j=1}^{n} \lambda_j \Big\langle \mathbf{v}_j, \sum_{i=1}^{n} \alpha_i(t) v_i \Big\rangle^{k-1} \mathbf{v}_j = \sum_{j=1}^{n} \lambda_j \alpha_j(t)^{k-1} \mathbf{v}_j$$

Therefore, the TVPDS can be simplified into n one-dimensional nonlinear dynamical systems, i.e.,

$$\dot{\alpha}_j(t) = \lambda_j \alpha_j(t)^{k-1}$$

for $j = 1, 2, \ldots, n$. By the method of separation of variables, the solutions of $\alpha_j(t)$ with the initial conditions $\alpha_j(0) = c_j$ can be written as

$$\alpha_j(t) = \left(1 - (k-2)\lambda_j c_j^{k-2} t\right)^{-\frac{1}{k-2}}.$$

Hence, the result follows immediately. $\qquad\square$

Clearly, if $\lambda_j c_j^{k-2} > 0$ for some j, $\alpha_j(t)$ will have singularity points at $t = \frac{1}{(k-2)\lambda_j c_j^{k-2}}$. Thus, the domain of the solution (4.3) is given by

$$D = \left[0, \min_{S} \frac{1}{(k-2)\lambda_j c_j^{k-2}}\right),$$

and the system blows up within a finite time. If $\lambda_j c_j^{k-2} \leq 0$ for all j, the domain of the solution (4.3) is $[0, \infty)$. Furthermore, when $k = 2$, the explicit solution (4.3) reduces to the classical solutions of LDSs, i.e.,

$$\lim_{k \to 2} \mathbf{x}(t) = \lim_{k \to 2} \sum_{j=1}^{n} \left(1 - (k-2)\lambda_j c_j^{k-2} t\right)^{-\frac{1}{k-2}} c_j \mathbf{v}_j$$

$$= \lim_{\tilde{k} \to \infty} \sum_{j=1}^{n} \left(1 + \frac{\lambda_j t}{\tilde{k}}\right)^{\tilde{k}} c_j \mathbf{v}_j = \sum_{j=1}^{n} \exp\{\lambda_j t\} c_j \mathbf{v}_j,$$

where λ_j become the eigenvalues of the dynamic matrix with the corresponding eigenvectors $\mathbf{v}_j$.

4.2.2 Stability

The stability properties of odeco TVPDSs can be obtained directly from the explicit solution formula (4.3). First, it is essential to understand the number of equilibrium points of an odeco TVPDS.

Proposition 4.2 ([2]) *The odeco TVDPS (4.1) has a unique equilibrium point at the origin if $\lambda_j \neq 0$ for all $j = 1, 2, \ldots, n$ and has infinitely many equilibrium points if $\lambda_j = 0$ for some $j = 1, 2, \ldots, n$, where λ_j are the Z-eigenvalues from the orthogonal decomposition of A.*

Proof Suppose that $\mathbf{x}_e = \sum_{j=1}^{n} c_j \mathbf{v}_j$ is an equilibrium point ($\mathbf{v}_j$ are the corresponding Z-eignvectors). Based on the property of tensor vector products, it can be shown that

$$\mathsf{A}\mathbf{x}_e^{k-1} = \sum_{j=1}^{n} \lambda_j c_j^{k-1} \mathbf{v}_j = \mathbf{0}.$$

Therefore, the result follows immediately. $\qquad\square$

The number of equilibrium points of an odeco TVPDS is similar to that of LDSs. Thus, it is necessary to exclusively focus on the equilibrium point at the origin because the behaviors of other equilibrium points will be the same if $\lambda_j = 0$ for some j. Analogous to linear stability, the equilibrium point $\mathbf{x}_e = \mathbf{0}$ of an odeco TVPDS is called stable if $\|\mathbf{x}(t)\| \leq \beta \|\mathbf{x}_0\|$ given the initial condition $\mathbf{x}_0$ and some $\beta > 0$, asymptotically stable if $\lim_{t\to\infty} \|\mathbf{x}(t)\| = 0$, and unstable if $\lim_{t\to s} \|\mathbf{x}(t)\| = \infty$ for $s > 0$. This instability definition is also known as finite-time blow-up in the nonlinear system literature [21].

Proposition 4.3 ([2]) *Suppose that the initial condition* $\mathbf{x}_0 = \sum_{j=1}^{n} c_j \mathbf{v}_j$. *The equilibrium point* $\mathbf{x}_e = \mathbf{0}$ *of the odeco TVPDS (4.1) is:*

- *stable if and only if* $\lambda_j c_j^{k-1} \leq 0$ *for all* $j = 1, 2, \ldots, n$;
- *asymptotically stable if and only if* $\lambda_j c_j^{k-1} < 0$ *for all* $j = 1, 2, \ldots, n$;
- *unstable if and only if* $\lambda_j c_j^{k-1} > 0$ *for some* $j = 1, 2, \ldots, n$,

where λ_j *are the Z-eigenvalues from the orthogonal decomposition of* A *with the corresponding Z-eigenvector* $\mathbf{v}_j$.

Proof The results follow immediately based on the explicit solution formula (4.3) of odeco TVPDSs. $\qquad\square$

When k is even, the results can be further simplified.

Corollary 4.1 ([2]) *Suppose that k is even. The equilibrium point* $\mathbf{x}_e = \mathbf{0}$ *of the odeco TVPDS (4.1) is:*

- *stable if and only if* $\lambda_j \leq 0$ *for all* $j = 1, 2, \ldots, n$;
- *asymptotically stable if and only if* $\lambda_j < 0$ *for all* $j = 1, 2, \ldots, n$;
- *unstable if and only if* $\lambda_j > 0$ *for some* $j = 1, 2, \ldots, n$,

where λ_j *are the Z-eigenvalues from the orthogonal decomposition of* A.

Proof The results follow immediately from Proposition 4.3. $\qquad\square$

If $k = 2$, the above conditions reduce to the classical linear stability conditions. In addition, the region of attraction of the odeco TVPDS (4.1) can be obtained based on Proposition 4.3, which is computed as

$$R = \left\{ \mathbf{x} : \lambda_j c_j^{k-2} < 0 \text{ where } \mathbf{x} = \sum_{j=1}^{n} c_j \mathbf{v}_j \right\}. \tag{4.4}$$

Additionally, as previously mentioned, U-eigenvalues are the generalization of Z-eigenvalues for even-order supersymmetric tensors. Therefore, U-eigenvalues are utilized to approximate the maximum Z-eigenvalue of an even-order tensor.

Lemma 4.1 ([19]) *Suppose that* $A \in \mathbb{R}^{(n \times n) \times \overset{k}{\cdots} \times (n \times n)}$ *is an even-order (i.e., 2kth-order) supersymmetric tensor. The largest Z-eigenvalue* $\lambda_{\max}$ *of* A *is upper bounded by the largest U-eigenvalue* $\mu_{\max}$ *of* A.

Proof The result can be obtained directly from the two optimization problems (1.20) and (1.22). □

Corollary 4.2 ([2]) *Suppose that* k *is even. The equilibrium point* $\mathbf{x}_e = \mathbf{0}$ *of the odeco TVPDS (4.1) is:*

- *stable if* $\mu_{\max} \leq 0$;
- *asymptotically stable if* $\mu_{\max} < 0$,

where $\mu_{\max}$ *is the largest U-eigenvalue of* A.

Proof The Z-eigenvalues from the orthogonal decomposition of A do not include all the Z-eigenvalues of A. Therefore, $\lambda_1 \leq \lambda_{\max} \leq \mu_{\max}$. Then the results follow immediately from Corollary 4.1. □

4.2.3　Discrete-Time Case

Similarly to continuous-time odeco TVPDSs, explicit solutions and stability properties can also be derived for discrete-time odeco TVPDSs, i.e.,

$$\mathbf{x}_{t+1} = A\mathbf{x}_t^{k-1}, \tag{4.5}$$

where $A \in \mathbb{R}^{n \times n \times \overset{k}{\cdots} \times n}$ is odeco.

Proposition 4.4 ([19]) *Suppose $A \in \mathbb{R}^{n \times n \times \overset{k}{\cdots} \times n}$ is odeco as in Proposition 4.1. Let the initial condition $x_0 = \sum_{j=1}^{n} c_j v_j$. Given the initial condition x_0, the explicit solution of the discrete-time odeco TVPDS (4.5) at time q, can be computed as*

$$x_q = \sum_{j=1}^{n} \lambda_j^{\alpha} c_j^{\beta} v_j, \tag{4.6}$$

where $\alpha = \sum_{j=0}^{q-1}(k-1)^j = \frac{(k-1)^q - 1}{k-2}$ and $\beta = (k-1)^q$.

Proposition 4.5 ([19]) *Suppose that the initial condition $x_0 = \sum_{j=1}^{n} c_j v_j$. The equilibrium point $x_e = 0$ of the discrete-time odeco TVPDS (4.5) is:*

- *stable if and only if $|c_j \lambda_j^{\frac{1}{k-2}}| \le 1$ for all $j = 1, 2, \ldots, n$;*

- *asymptotically stable if and only if $|c_j \lambda_j^{\frac{1}{k-2}}| < 1$ for all $j = 1, 2, \ldots, n$;*

- *unstable if and only if $|c_j \lambda_j^{\frac{1}{k-2}}| > 1$ for some $j = 1, 2, \ldots, n$,*

where λ_j are the Z-eigenvalues from the orthogonal decomposition of A with the corresponding Z-eigenvectors v_j.

Propositions 4.4 and 4.5 can be proven in a similar manner to Propositions 4.1 and 4.3, respectively, by exploiting the orthogonal structure of the dynamic tensor.

4.2.4 Constant Control

Consider the odeco TVPDS (4.1) with constant control inputs, i.e.,

$$\dot{\mathbf{x}}(t) = A\mathbf{x}(t)^{k-1} + \mathbf{b}, \tag{4.7}$$

where $\mathbf{b} \in \mathbb{R}^n$ is a constant input vector. The constant control problem arises in numerous dynamical systems and control applications. For instance, in population dynamics, constant inputs can represent migration or supply rates for different population species [22, 23]. Additionally, constant inputs have been effectively used in model predictive control for stable plants based on linear models [24, 25]. The complete solution of the odeco TVPDS with constant control inputs (4.7) can be obtained implicitly using the Gauss hypergeometric functions.

Proposition 4.6 ([2]) *Suppose that $x(t) = \sum_{j=1}^{n} \alpha_j(t) v_j$ with initial condition $\alpha_j(0) = c_j$. For the odeco TVPDS with constant control inputs (4.7), the coefficient functions $\alpha_j(t)$ can be solved implicitly by*

$$t = -\frac{g\left(\frac{k-2}{k-1}, -\frac{\tilde{b}_j}{\lambda_r j \alpha_j(t)^{k-1}}\right)}{(k-2)\lambda_j \alpha_j(t)^{k-2}} + \frac{g\left(\frac{k-2}{k-1}, -\frac{\tilde{b}_j}{\lambda_j c_j^{k-1}}\right)}{(k-2)\lambda_j c_j^{k-2}}, \tag{4.8}$$

where $g(\cdot, \cdot)$ is a specified Gauss hypergeometric function [26, 27] defined as

$$g(a, z) = {}_2F_1(1, a; a+1; z) = a \sum_{m=0}^{\infty} \frac{z^m}{a+m},$$

λ_j are the Z-eigenvalues from the orthogonal decomposition of A with the corresponding Z-eigenvectors v_j, and $\tilde{b}_j$ are the jth entries of the vector $V^{\top} b$ (V contains all the vector of v_j).

Proof Similar to Proposition 4.1, the TVPDS with constant control input (4.7) can be rewritten into n one-dimensional systems, i.e.,

$$\dot{\alpha}_j(t) = \lambda_j \alpha_j(t)^{k-1} + \tilde{b}_j.$$

The one-dimensional differential equations can be solved implicitly by using the Gauss hypergeometric functions. By the method of separation of variables, the solutions of $\alpha_j(t)$ with initial condition $\alpha_j(0) = c_j$ can be solved as (4.8). $\qquad\square$

The one-dimensional differential equations in the proof above fall under a specific form of Chini's equations [28]. When $k = 3$, they are known as Riccati equations [29]. Upon obtaining the implicit equations (4.8), one can employ any nonlinear solver to determine the values of $\alpha_j(t)$ for a specific time t, allowing us to recover the complete solution of the odeco TVPDS with constant control inputs (4.7). The stability properties follow as well. While $g(a, z)$ is only defined for $|z| < 1$, it can be analytically continued along any path in the complex plane that avoids the branch points of one and infinity [30].

4.2.5   Generalization

This subsection aims to explore the stability properties of more general TVPDSs with almost symmetric dynamic tensors (i.e., arbitrary homogeneous polynomial dynamical systems) using linear transformations. Since A is almost symmetric, a CPD of A can be computed as

$$A = \sum_{j=1}^{r} \overbrace{v_j \circ v_j \circ \cdots \circ v_j}^{k-1} \circ u_j. \tag{4.9}$$

The weights λ_j are omitted here. The objective is to create a linear transformation $\mathbf{P} \in \mathbb{R}^{n \times n}$ that can convert the TVPDS (4.1) with an almost symmetric dynamic tensor A into an odeco TVPDS.

Proposition 4.7 ([2]) *Suppose that* $\mathsf{A} \in \mathbb{R}^{n \times n \times \overset{k}{\cdots} \times n}$ *has a CPD of the form (4.9) with* $r = n$. *If there exist an invertible matrix* $\mathbf{P} \in \mathbb{R}^{n \times n}$ *and a diagonal matrix* $\mathbf{\Lambda} \in \mathbb{R}^{n \times n}$ *such that* $\mathbf{P}^{\top} \mathbf{V} = \mathbf{P}^{-1} \mathbf{U} \mathbf{\Lambda}^{-1}$ *and* $\mathbf{P}^{\top} \mathbf{V}$ *is an orthogonal matrix where* $\mathbf{V} \in \mathbb{R}^{n \times n}$ *and* $\mathbf{U} \in \mathbb{R}^{n \times n}$ *contain all the vectors of* $\mathbf{v}_r$ *and* $\mathbf{u}_r$, *respectively, the TVPDS (4.1) can be transformed to an odeco TVPDS with the transformation matrix* $\mathbf{P}$.

Proof Suppose that $\mathbf{y}(t) = \mathbf{P}^{-1} \mathbf{x}(t)$. Based on the property of tensor vector products, it can be shown that

$$\dot{\mathbf{y}}(t) = \mathbf{P}^{-1} \left(\mathsf{A} \mathbf{x}(t)^{k-1} \right)$$

$$= \mathbf{P}^{-1} \left(\left(\sum_{j=1}^{n} \overbrace{\mathbf{v}_j \circ \mathbf{v}_j \circ \cdots \circ \mathbf{v}_j}^{k-1} \circ \mathbf{u}_j \right) \left(\mathbf{P} \mathbf{y}(t) \right)^{k-1} \right)$$

$$= \left(\sum_{j=1}^{n} \overbrace{\mathbf{P}^{\top} \mathbf{v}_j \circ \mathbf{P}^{\top} \mathbf{v}_j \circ \cdots \circ \mathbf{P}^{\top} \mathbf{v}_j}^{k-1} \circ \mathbf{P}^{-1} \mathbf{u}_j \right) \mathbf{y}(t)^{k-1}.$$

Hence, if $\mathbf{P}^{\top} \mathbf{V} = \mathbf{P}^{-1} \mathbf{U} \mathbf{\Lambda}^{-1}$, the above differential equation can be rewritten as

$$\dot{\mathbf{y}}(t) = \left(\sum_{j=1}^{n} \overbrace{\mathbf{P}^{\top} \mathbf{v}_j \circ \mathbf{P}^{\top} \mathbf{v}_j \circ \cdots \circ \mathbf{P}^{\top} \mathbf{v}_j}^{k-1} \circ \lambda_r \mathbf{P}^{\top} \mathbf{v}_j \right) \mathbf{y}(t)^{k-1},$$

where λ_j are the $j$the diagonal entries of $\mathbf{\Lambda}$. If $\mathbf{P}^{\top} \mathbf{V}$ is an orthogonal matrix, the transformed system is an odeco TVPDS. $\qquad\square$

Corollary 4.3 ([2]) *Suppose that* $\mathsf{A} \in \mathbb{R}^{n \times n \times \overset{k}{\cdots} \times n}$ *has a CPD of the form (4.9) with* $r = n$. *If there exist* λ_j *such that* $\mathbf{w}_j = \lambda^{-1} \mathbf{u}_j$ *where* $\mathbf{w}_j$ *are the jth column of the matrix* $(\mathbf{V}^{-1})^{\top}$, *the TVPDS (4.1) can be transformed to an odeco TVPDS.*

Proof The result can be obtained immediately by combining the two conditions stated in Proposition 4.7. $\qquad\square$

One can leverage the MATLAB toolbox TensorLab [31] to determine whether a general TVPDS can be transformed to an odeco TVPDS, as outlined in Algorithm 5. In fact, this algorithm is applicable to any TVPDS, regardless of the CP rank of the dynamic tensor, and can identify an approximation (controlled by the error parameter ϵ) that can be transformed to

---

Algorithm 5 Determining if a general TPVDS (with almost symmetric dynamic tensor A) can be transformed to an odeco TVPDS. This algorithm was adapted from [2] with permission.

1: Given a general TPVDS of the form (4.1) and a threshold ϵ (default: $\epsilon = 10^{-14}$)
2: Create a CPD model

```
model = struct
```

3: Randomly initialize a variable $\mathbf{R} \in \mathbb{R}^{n \times n}$ and a weight vector $\lambda \in \mathbb{R}^n$

```
model.variables.R = randn(n,n)
model.variables.lambda = randn(1,n)
```

4: Impose the condition on the structure of the factor matrices such that $\mathbf{V} = \mathbf{R}$ and $\mathbf{V}^{(f)} = (\mathbf{R}^{-1})^{\top}$

```
model.factors.lambda = 'lambda'
model.factors.V = 'R'
model.factor.U = {'R',@(z,task)struct_invtransp(z, task)}
```

5: Compute a CPD of the dynamic tensor A based on the imposed conditions

```
model.factorizations.symm.data = A
model.factorizations.cpd ={'V',...,'V','U','lambda'}
cpd = sdf_nls(model)
```

6: Use the obtained factor matrices $\mathbf{V}$ and $\mathbf{U}$ along with the weight vector λ to build the estimated tensor $\hat{\mathbf{A}}$

```
lambda = cpd.factors.lambda
V = cpd.factors.V
U = cpd.factors.U
Ahat = cpdgen({V,...,V,U,lambda})
```

7: **if** $\|A - \hat{A}\| < \epsilon$ **then**

8:     The TVPDS with almost symmetric dynamic tensor A can be transformed to an odeco TVPDS

9: **end if**

---

an odeco TVPDS. Once establishing that a TVPDS (or its approximation) can be transformed to an odeco TVPDS, the linear transformation $\mathbf{P}$ can be readily computed using the formula $\mathbf{P} = (\mathbf{V}^{-1}\mathbf{W})^{\top}$, where $\mathbf{W} \in \mathbb{R}^{n \times n}$ is an arbitrary orthogonal matrix. The stability results can also be obtained immediately.

4.3   System-Theoretic Properties: Controllability

The controllability of polynomial dynamical systems has been an active area of research since the 1970s and 1980s [32–34]. Notably et al. [32] achieved a significant breakthrough by demonstrating the strong controllability of homogeneous polynomial dynamical systems (i.e., TVPDSs) with linear control inputs. This section focuses on harnessing the power of tensor algebra to establish a controllability/accessibility result for TVPDSs and explore its applications in the context of hypergraphs.

4.3.1 **Controllability**

Consider the following TVPDS with linear control inputs:

$$\dot{\mathbf{x}}(t) = \mathbf{A}\mathbf{x}(t)^{k-1} + \mathbf{B}\mathbf{u}(t), \tag{4.10}$$

where $\mathbf{B} \in \mathbb{R}^{n \times s}$ is the control matrix and $\mathbf{u}(t) \in \mathbb{R}^{s}$ is the control inputs. For simplicity, assume that the dynamic tensor $\mathbf{A}$ is supersymmetric. Generalization to almost symmetric dynamic tensors is straightforward. First, the controllability result for homogeneous polynomial dynamical systems is state below according to [32].

Definition 4.1 ([32]) A dynamical system is said to be strongly controllable if there exists a choice of control inputs that can drive it from any initial state to any target state at any time.

Theorem 4.1 ([32]) *Suppose that f is an n-dimensional homogeneous polynomial dynamical system of odd degree and $b_j \in \mathbb{R}^n$. The following control system:*

$$\dot{x}(t) = f(x(t)) + \sum_{j=1}^{s} b_j u_j(t), \tag{4.11}$$

is strongly controllable if and only if the rank of the Lie algebra spanned by the set of vector fields $\{f, b_1, b_2, \ldots, b_s\}$ is n at all points in $\mathbb{R}^n$.

It is known that the Lie algebra is of full rank at all points in $\mathbb{R}^n$ if and only if it is of full rank at the origin. Therefore, one can evaluate the Lie algebra of $\mathbf{f}, \mathbf{b}_1, \mathbf{b}_2, \ldots, \mathbf{b}_s$ at the origin by computing their recursive Lie brackets. The Lie bracket of two vector fields $\mathbf{f}$ and $\mathbf{g}$ at a point $\mathbf{x}$ can be easily computed as

$$[\mathbf{f}, \mathbf{g}]_{\mathbf{x}} = \nabla \mathbf{g}(\mathbf{x})\mathbf{f}(\mathbf{x}) - \nabla \mathbf{f}(\mathbf{x})\mathbf{g}(\mathbf{x}), \tag{4.12}$$

where ∇ is the gradient operator. Theorem 4.1 can be summarized as a generalized Kalman's rank condition for TVPDSs in the tensor form.

Definition 4.2 ([1]) Given the TVPDS with linear control inputs (4.10), define $\mathscr{C}_0$ as the linear span of $\{\mathbf{b}_1, \mathbf{b}_2, \ldots, \mathbf{b}_m\}$ where $\mathbf{B} = \begin{bmatrix} \mathbf{b}_1 & \mathbf{b}_2 & \ldots & \mathbf{b}_s \end{bmatrix} \in \mathbb{R}^{n \times s}$ and compute $\mathscr{C}_q$ inductively as the linear span of

$$\mathscr{C}_{q-1} \cup \{\mathbf{A}\mathbf{v}_1 \mathbf{v}_2 \cdots \mathbf{v}_{k-1} | \mathbf{v}_l \in \mathscr{C}_{q-1}\}. \tag{4.13}$$

for each integer $q \geq 1$. Denote the subspace $\mathscr{C}(\mathbf{A}, \mathbf{B}) = \cup_{q \geq 0} \mathscr{C}_q$.

Proposition 4.8 ([1]) *Suppose that k is even. The TVPDS with linear control inputs (4.10) is strongly controllable if and only if the subspace $\mathscr{C}(A, B)$ spans $\mathbb{R}^n$.*

Proof Without loss of generality, assume that $s = 1$. The recursive Lie brackets of $\{A, \mathbf{b}_1\}$ at the origin can be computed as

$$[\mathbf{b}_1, A\mathbf{x}^{k-1}]_0 = \left(\frac{d}{d\mathbf{x}}\Big|_{\mathbf{x}=0} A\mathbf{x}^{k-1}\right)\mathbf{b}_1 = \mathbf{0},$$

$$[\mathbf{b}_1, [\mathbf{b}_1, A\mathbf{x}^{k-1}]]_0 = \left(\frac{d}{d\mathbf{x}}\Big|_{\mathbf{x}=0} A\mathbf{x}^{k-2}\mathbf{b}_1\right)\mathbf{b}_1 = \mathbf{0},$$

$$\vdots$$

$$[\mathbf{b}_1, [\cdots, [[\mathbf{b}_1, A\mathbf{x}^{k-1}]]]]_0 = \left(\frac{d}{d\mathbf{x}}\Big|_{\mathbf{x}=0} A\mathbf{x}\mathbf{b}_1^{k-2}\right)\mathbf{b}_1 = A\mathbf{b}_1^{k-1}.$$

Here, the scalar terms are ignored during the gradient computation. One can repeat the process for the brackets $[A\mathbf{b}_1^{k-1}, A\mathbf{x}^{k-1}]$, $[A\mathbf{b}_1^{k-1}, A\mathbf{x}^{k-2}\mathbf{b}_1]$, $\ldots$, $[A\mathbf{b}_1^{k-1}, A\mathbf{x}\mathbf{b}_1^{k-2}]$. After the qth iteration, the subspace $\mathscr{C}(A, B)$ contains all the Lie brackets of the vector fields $A\mathbf{x}^{k-1}, \mathbf{b}_1$ at the origin. Hence, the desired result follows immediately from Theorem 4.1. $\qquad\square$

An integer $q \leq n$ can be determined such that $\mathscr{C}(A, B) = \mathscr{C}_q$, as $\mathscr{C}(A, B)$ is a finite-dimensional vector space. Additionally, the matrix whose columns consist of the vectors from $\mathscr{C}(A, B)$ can be considered as the controllability matrix of the TVPDS with linear control inputs (4.10). When $k = 2$ and $q = n - 1$, Proposition 4.8 reduces to the famous Kalman's rank condition for linear control systems. In lieu of using the tensor vector product, the controllability matrix (in a reduced form) can be efficiently computed using the Kronecker product and matrix SVD, as shown in Algorithm 6. It can be demonstrated that the column space of the controllability matrix obtained from Algorithm 6 is equivalent to $\mathscr{C}(A, B)$.

The controllability of the TVPDS with linear control inputs (4.10) remains an open problem for odd k [35, 36]. However, a weaker form of controllability, known as local accessibility, can be established based on the generalized Kalman's rank condition. Local accessibility only requires that reachable sets be nonempty open sets rather than the entire space $\mathbb{R}^n$ [37].

Definition 4.3 ([37]) A dynamical system is called locally accessible if for any initial state $\mathbf{x}_0 \in \mathbb{R}^n$ and $T > 0$, the reachable set $\mathscr{R}_T(\mathbf{x}_0) = \cup_{0 \leq t \leq T} \mathscr{R}(\mathbf{x}_0, t)$ contains a nonempty open set, where $\mathscr{R}(\mathbf{x}_0, t)$ is the set including all $\mathbf{x} \in \mathbb{R}^n$ for which the system can be driven from $\mathbf{x}_0$ to $\mathbf{x}$ at time t.

Algorithm 6 Computing the controllability matrix. This algorithm was adapted from [1] with permission.

1: Given a TVPDS (4.10) with supersymmetric dynamic tensor A and control matrix **B**
2: Compute the 1-mode matricization of A and denote it by $\mathbf{A}_{(1)} \in \mathbb{R}^{n \times n^{k-1}}$
3: Set $\mathbf{C} = \mathbf{B}$ and $j = 0$
4: **while** $j < n$ **do**

5:     Compute $\mathbf{T} = \mathbf{A}_{(1)}(\mathbf{C} \otimes \mathbf{C} \otimes \overset{k-1}{\cdots} \otimes \mathbf{C})$
6:     Set $\mathbf{C} = \begin{bmatrix} \mathbf{C} & \mathbf{T} \end{bmatrix}$
7:     Compute the economy-size SVD of **C**, i.e., $\mathbf{C} = \mathbf{U}\mathbf{S}\mathbf{V}^{\top}$
8:     Set $\mathbf{C} = \mathbf{U}$ and $j = j + 1$

9: **end while**
10: **return** The controllability matrix (in a reduced form) **C** of the TVPDS.

Proposition 4.9 ([1]) *The TVPDS with linear control inputs (4.10) is locally accessible if and only if the subspace* $\mathscr{C}(A, \boldsymbol{B})$ *spans* $\mathbb{R}^n$.

Proof According to nonlinear control theory, accessibility can be achieved when the span of the smallest Lie algebra of vector fields containing the drift and input vector fields is $\mathbb{R}^n$. For the TVPDS with linear control inputs (4.10), the smallest Lie algebra containing $A\mathbf{x}^{k-1}$ and $\mathbf{b}_1, \mathbf{b}_2, \ldots, \mathbf{b}_m$ at the origin is $\mathscr{C}(A, \mathbf{B})$. Therefore, the result follows immediately. □

4.3.2   Controllability of Hypergraphs

The controllability of graphs has been extensively studied since the 1970s [38–44]. Liu et al. [40, 44] modeled the dynamics of a graph using the LDS model, where the dynamic matrix is the adjacency matrix of the graph. They utilized the Kalman's rank condition to determine the minimum number of driver nodes required to control the entire graph. A hypergraph is a generalization of a graph in which its hyperedges can connect more than two nodes [8]. Hypergraphs are commonly used to model various systems such as metabolic reaction networks [45, 46], co-authorship networks [47, 48], email communication networks [49, 50], and protein-protein interaction networks [51, 52]. If every hyperedge of a hypergraph contains k nodes, the hypergraph is called a k-uniform hypergraph, as shown in Fig. 4.1B (Fig. 4.1A shows a non-uniform hypergraph). Uniform hypergraphs can be efficiently represented by tensors.

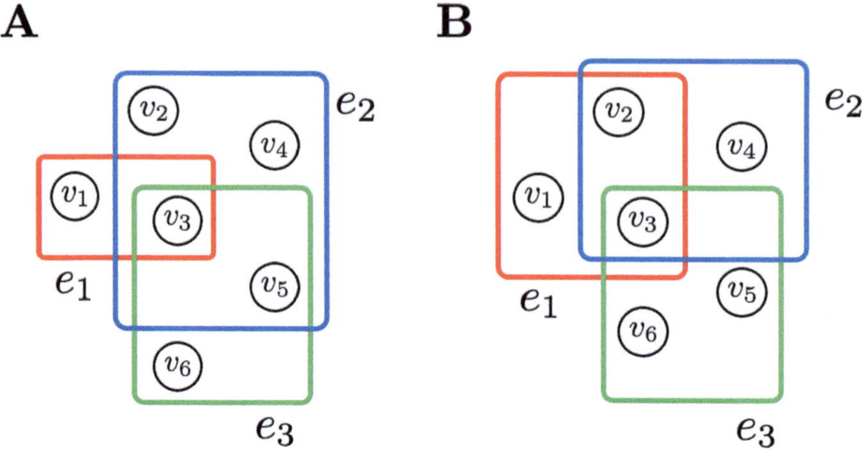

Fig. 4.1 Non-uniform hypergraphs versus uniform hypergraphs. **A** Non-uniform hypergraph with hyperedges $e_1 = \{v_1, v_3\}$, $e_2 = \{v_2, v_3, v_4, v_5\}$, $e_3 = \{v_3, v_5, v_6\}$. **B** Uniform hypergraph with hyperedges $e_1 = \{v_1, v_2, v_3\}$, $e_2 = \{v_2, v_3, v_4\}$, $e_3 = \{v_3, v_5, v_6\}$

Definition 4.4 ([53]) Let $\mathcal{H} = \{\mathcal{V}, \mathcal{E}\}$ be a k-uniform hypergraph with n nodes. The adjacency tensor $\mathsf{A} \in \mathbb{R}^{n \times n \times \overset{k}{\cdots} \times n}$ of $\mathcal{H}$, a kth order n-dimensional supersymmetric tensor, is defined as

$$\mathsf{A}_{j_1 j_2 \cdots j_k} = \begin{cases} \frac{1}{(k-1)!} & \text{if } (j_1, j_2, \ldots, j_k) \in \mathcal{E} \\ \\ 0, & \text{otherwise} \end{cases} . \tag{4.14}$$

The degree of node j of a uniform hypergraph can be computed as

$$d_j = \sum_{j_2=1}^{n} \sum_{j_3=1}^{n} \cdots \sum_{j_k=1}^{n} \mathsf{A}_{j j_2 j_3 \cdots j_k}. \tag{4.15}$$

If all nodes have the same degree d, $\mathcal{H}$ is called d-regular. If every k nodes are contained in one hyperedge, $\mathcal{H}$ is called complete.

The dynamics of a k-uniform hypergraph $\mathcal{H}$ with linear control can be described by the TVPDS (4.10), where the dynamic tensor is defined as the adjacency tensor of $\mathcal{H}$ (see Fig. 4.2B). This allows us to analyze the controllability of uniform hypergraphs using the generalized Kalman's rank condition, which aims to determine the minimum number of driver nodes necessary for achieving the controllability of the hypergraph, with each input being imposed at only one node. However, Proposition 4.8 shows that controllability can only be achieved for even uniform hypergraphs (Fig. 4.2C provides a simple example). Chen et al. [1] proved that the minimum number of driver nodes of even uniform hyperchains, hyperstars, and complete even uniform hypergraphs behaves similarly to those of chains,

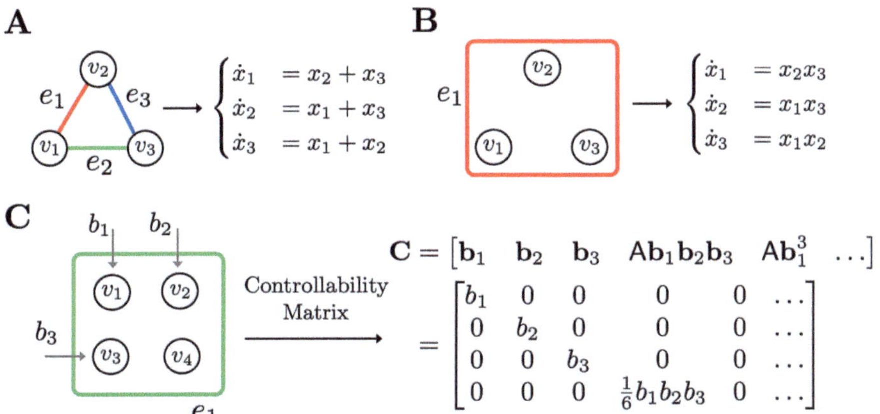

Fig. 4.2 Hypergraph dynamics and controllability matrix. **A** Standard graph with three nodes and three edges and its associated linear dynamics. **B** 3-uniform hypergraph with three nodes and one hyperedge and its associated nonlinear dynamics. **C** 4-uniform hypergraph with four nodes and one hyperedge and its corresponding controllability matrix with three inputs. This figure was redrawn from [1] with permission

stars, and complete graphs and reduce them as special cases for $k = 2$. Reader can refer to the work [1] for detailed definitions of hyperchains and hyperstars. Here, the result for complete even uniform hypergraphs is provided as follows.

Proposition 4.10 ([1]) *If $\mathcal{H}$ is a complete even uniform hypergraph with n nodes, then the minimum number of driver nodes of $\mathcal{H}$ is $n - 1$.*

The results for the minimum number of driver nodes in odd uniform hypergraphs are exactly the same as those for even uniform hypergraphs when considering hypergraph accessibility. Determining the minimum number of driver nodes (whether it is controllability or accessibility) for a uniform hypergraph is an NP-hard problem. Nevertheless, a heuristic algorithm (Algorithm 7) is proposed to estimate this number for uniform hypergraphs in [1]. The algorithm selects nodes that cause the maximum change in the rank of the reduced controllability matrix computed from Algorithm 6. Here, $\mathbf{C}_{\mathcal{D}}$ represents the controllability matrix computed from the inputs formed by the set $\mathcal{D}$ of driver nodes. For non-connected uniform hypergraphs, it is necessary to identify the connected components first. If the algorithm returns multiple optimal solutions, one can randomly choose one or use additional criteria, such as selecting the node with the highest degree, to break the tie. Numerical experiments have shown that Algorithm 7 can estimate the minimum number of driver nodes for medium-sized uniform hypergraphs with high accuracy and at a much lower computational cost compared to brute-force search.

Algorithm 7 Greedy driver node selection. This algorithm was adapted from [1] with permission.

1: Given the adjancency tensor $\mathsf{A} \in \mathbb{R}^{(n \times n) \times \overset{k}{\cdots} \times (n \times n)}$ of an even uniform hypergraph $\mathcal{H}$
2: Let $\mathcal{S} = \{1, 2, \ldots, n\}$ and $\mathcal{D} = \emptyset$
3: **while** rank$(\mathbf{C}_{\mathcal{D}}) < n$ **do**

4:     **for** $s \in \mathcal{S} \setminus \mathcal{D}$ **do**

5:        Compute $\Delta(s) = \text{rank}(\mathbf{C}_{\mathcal{D} \cup \{s\}}) - \text{rank}(\mathbf{C}_{\mathcal{D}})$ using Algorithm 6

6:     **end for**
7:     Set $s^* = \text{argmax}_{s \in \mathcal{S} \setminus \mathcal{D}} \Delta(s)$
8:     Set $\mathcal{D} = \mathcal{D} \cup \{s^*\}$

9: **end while**
10: **return** A subset of driver nodes $\mathcal{D}$.

4.4   System-Theoretic Property: Observability

The nonlinearity of TVPDSs poses challenges in extending the framework of controllability to observability. In fact, there are no strong results for the observability of polynomial dynamical systems. Therefore, one has to rely on the results of weak observability based on nonlinear systems theory [54, 55].

4.4.1 Observability

Consider the following TVPDS with linear output:

$$\begin{cases} \dot{\mathbf{x}}(t) = \mathsf{A}\mathbf{x}(t)^{k-1} \\ \mathbf{y}(t) = \mathbf{C}\mathbf{x}(t) \end{cases}, \tag{4.16}$$

where $\mathbf{C} \in \mathbb{R}^{m \times n}$ is the output matrix and $\mathbf{y}(t) \in \mathbb{R}^m$ is the output. For simplicity, assume that the dynamic tensor A is supersymmetric. First, the observability result for nonlinear systems is state below according to [54]. Consider the following n-dimensional nonlinear system with inputs and outputs:

$$\begin{cases} \dot{\mathbf{x}}(t) = \mathbf{f}(\mathbf{x}(t), \mathbf{u}(t)) \\ \mathbf{y}(t) = \mathbf{g}(\mathbf{x}(t)) \end{cases}, \tag{4.17}$$

where $\mathbf{f}$ and $\mathbf{g} = \begin{bmatrix} g_1 & g_2 & \cdots & g_m \end{bmatrix}^\top$ are analytic vector fields.

Definition 4.5 ([54]) Let U be a subset of $\mathbb{R}^n$. Two points $\mathbf{x}_0$ and $\mathbf{x}_1$ in U are called U-indistinguishable if for every control input $\mathbf{u}(t)$ over $[0, T]$, whose solutions $\mathbf{x}_0(t)$ and $\mathbf{x}_1(t)$ from $\mathbf{x}_0$ and $\mathbf{x}_1$ lie in U, fails to distinguish between $\mathbf{x}_0$ and $\mathbf{x}_1$.

Definition 4.6 ([54]) The nonlinear system (4.17) is called locally weakly observable at $\mathbf{x} \in \mathbb{R}^n$ if and only if $\mathbf{x}$ has an open neighborhood U such that for every open neighborhood V of $\mathbf{x}$ contained in U, all points in V are U-indistinguishable from $\mathbf{x}$.

Theorem 4.2 ([54]) *The system is locally weakly observable at* $\mathbf{x} \in \mathbb{R}^n$ *if and only if the observability matrix defined as*

$$O(x) = \nabla \begin{bmatrix} L_f^0 g(x) & L_f^1 g(x) & \cdots & L_f^q g(x) \end{bmatrix}^\top \tag{4.18}$$

has rank n for some integer $q > 0$.

The notation $L_{\mathbf{f}}\mathbf{g}(\mathbf{x})$ represents the Lie derivative of $\mathbf{g}$ along $\mathbf{f}$ defined as

$$L_{\mathbf{f}}\mathbf{g}(\mathbf{x}) = \begin{bmatrix} \langle \mathbf{f}, \nabla g_1(\mathbf{x}) \rangle & \langle \mathbf{f}, \nabla g_2(\mathbf{x}) \rangle & \cdots & \langle \mathbf{f}, \nabla g_m(\mathbf{x}) \rangle \end{bmatrix}^\top ,$$

and $L_{\mathbf{f}}^j g(\mathbf{x})$ denotes the jth Lie derivative. In general, the value of q is set to n for analytic systems, and the generic rank condition of the observability matrix (4.18) can be checked using symbolic computation [56]. Furthermore, Theorem 4.2 can be simplified for TVPDSs using tensor algebra, similar to the framework of controllability. For convenience, define the following:

$$\mathbf{B}_p = \sum_{j=1}^{pk-k-2p+3} \overbrace{\mathbf{I} \otimes \mathbf{I} \otimes \cdots \otimes \mathbf{I}}^{j-1} \otimes \mathbf{A}_{(1)} \otimes \overbrace{\mathbf{I} \otimes \mathbf{I} \otimes \cdots \otimes \mathbf{I}}^{pk-k-2p+3-j},$$

where $\mathbf{A}_{(1)}$ is the 1-mode matricization of A.

Proposition 4.11 *([20]) The TVPDS with linear outputs (4.16) is locally weakly observable at* $\mathbf{x} \in \mathbb{R}^n$ *if and only if the observability matrix defined as*

$$O(x) = \nabla \begin{bmatrix} CJ_0(x) & CJ_1(x) & \cdots & CJ_n(x) \end{bmatrix}^\top , \tag{4.19}$$

where

$$J_0(x) = x,$$

$$J_1(x) = A_{(1)}(x \otimes x \otimes \overset{k-1}{\cdots} \otimes x),$$

$$J_p(x) = A_{(1)} B_2 B_3 \cdots B_p (x \otimes x \otimes \overset{pk-2p+1}{\cdots\cdots} \otimes x),$$

has rank n.

A detailed algorithm for computing the observability matrix can be found in [20]. When $k = 2$, Proposition 4.11 reduces to the Kalman's rank condition for the observability of LDSs.

4.4.2   Observability of Hypergraphs

The framework for hypergraph observability can be formulated similarly to that of hypergraph controllability. For a given uniform hypergraph, the dynamic tensor A in the TVPDS with linear outputs (4.16) is set to the adjacency tensor of the hypergraph. Each row in the output matrix **C** only contains one non-zero entry, similarly to the control matrix **B**. In other words, each output is imposed at only one node. The objective is to determine the minimum number of observable nodes necessary for achieving hypergraph observability. Algorithm 7 can be adapted to estimate the minimum number of observable nodes of a uniform hypergraph by simply replacing the controllability matrix with the observability matrix [20].

4.5   Applications

4.5.1   Synthetic Data: Stability

This example was adapted from [2]. Given a two-dimensional odeco TVPDS of the form (4.1) with

$$A_{::11} = \begin{bmatrix} -1.2593 & 0.5543 \\ 0.5543 & -0.5185 \end{bmatrix} \quad A_{::12} = \begin{bmatrix} 0.5543 & -0.5185 \\ -0.5185 & -0.1386 \end{bmatrix}$$

$$A_{::21} = \begin{bmatrix} 0.5543 & -0.5185 \\ -0.5185 & -0.1386 \end{bmatrix} \quad A_{::22} = \begin{bmatrix} -0.5185 & -0.1386 \\ -0.1386 & -0.7037 \end{bmatrix},$$

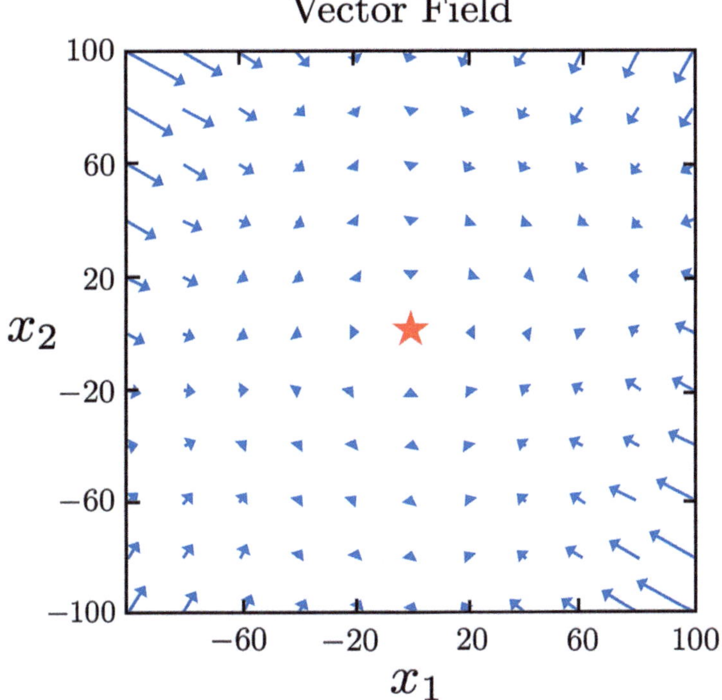

Fig. 4.3 Vector field plot of the odeco TVPDS. This figure was redrawn from [2] with permission

the orthogonal decomposition of A has two Z-eigenvalues, which are -1 and -2. Hence, according to Corollary 4.1, the TVPDS is globally asymptotically stable. Figure 4.3 shows the vector field of the system, where the origin is a sink.

4.5.2 Synthetic Data: Stability with Constant Control

This example was adapted from [2]. Given a three-dimensional homogeneous polynomial dynamical system of degree three with constant control inputs

$$\begin{cases} \dot{x}_1 = -x_1^3 - 3x_1^2 x_2 - 3x_1 x_2^2 + 2 \\ \dot{x}_2 = -x_2^3 + 2 \\ \dot{x}_3 = -x_3^3 - 3x_1^2 x_3 - 3x_1 x_3^2 - 3x_2^2 x_3 - 3x_2 x_3^2 - 6x_1 x_2 x_3 + 2 \end{cases},$$

it can be rewritten as a TVPDS of the form (4.7) with an almost symmetric A and $\mathbf{b} = \begin{bmatrix} 2 & 2 & 2 \end{bmatrix}^\top$. Algorithm 5 can be employed to transform the system into an odeco TVPDS with constant control. Assuming the standard basis is used to construct the odeco tensor (with Z-eigenvalues equal to -1 in the orthogonal decomposition), the coefficient functions $\alpha_j(t)$ can be solved by applying Proposition 4.6, i.e.,

$$
\begin{cases}
t = \dfrac{g(2/3,4/\alpha_1^3)}{2\alpha_1^2} - \dfrac{g(2/3,4/c_1^3)}{2c_1^2} \\[2mm]
t = \dfrac{g(2/3,2/\alpha_2^3)}{2\alpha_2^2} - \dfrac{g(2/3,2/c_2)}{2c_2^2} \\[2mm]
t = \dfrac{g(2/3,6/\alpha_3^3)}{2\alpha_3^2} - \dfrac{g(2/3,6/c_3^3)}{2c_3^2}
\end{cases} ,
$$

with $\alpha_r(0) = c_j$ for all $j = 1, 2, 3$. It can be proven that the three Gauss hypergeometric functions have vertical asymptotes at $c_1 = \sqrt[3]{4}$, $c_2 = \sqrt[3]{2}$, and $c_3 = \sqrt[3]{6}$, respectively. This implies that the odeco TVPDS with constant control is globally asymptotically stable at the equilibrium point. Consequently, the original polynomial dynamical system is also globally asymptotically stable at the equilibrium point.

4.5.3   Mouse Neuron Endomicroscopy Dataset

This example was adapted from [1]. The mouse endomicroscopy dataset comprises imaging videos capturing neuronal activities in a mouse hypothalamus during 10-minute periods of feeding, fasting, and re-feeding [57]. The objective is to differentiate the three phases quantitatively using the minimum number of driver nodes. Intuitively, a high minimum number of driver nodes of a hypergraph indicates that it will take more effort to control the hypergraph. First, a 3-uniform hypergraph is constructed based on the multi-correlation of the neuronal activities for each phase. The multi-correlation among three neurons is defined as

$$
\rho = \sqrt{1 - \det(\mathbf{R})}, \tag{4.20}
$$

where $\mathbf{R} \in \mathbb{R}^{3\times 3}$ is the correlation matrix of three neuronal activity levels [58]. A hyperedge is built when the multi-correlation of three neurons exceeds a prescribed threshold. Figure 4.4A–C. show the three 3-uniform hypergraphs constructed using the multi-correlation of the neuronal activities of a mouse under the feeding, fasting, and re-feeding phases. It is evident that the minimum number of driver nodes varies between the three phases. The fasting phase requires more neurons to be controlled due to fewer interactions, while the minimum number of driver nodes is greatly reduced due to an outburst of neuronal activities during the re-feeding phase, see Fig. 4.4D. On the other hand, the minimum number of driver nodes computed from the graph model fails to capture the changes in neuronal activity, see Fig. 4.4D.

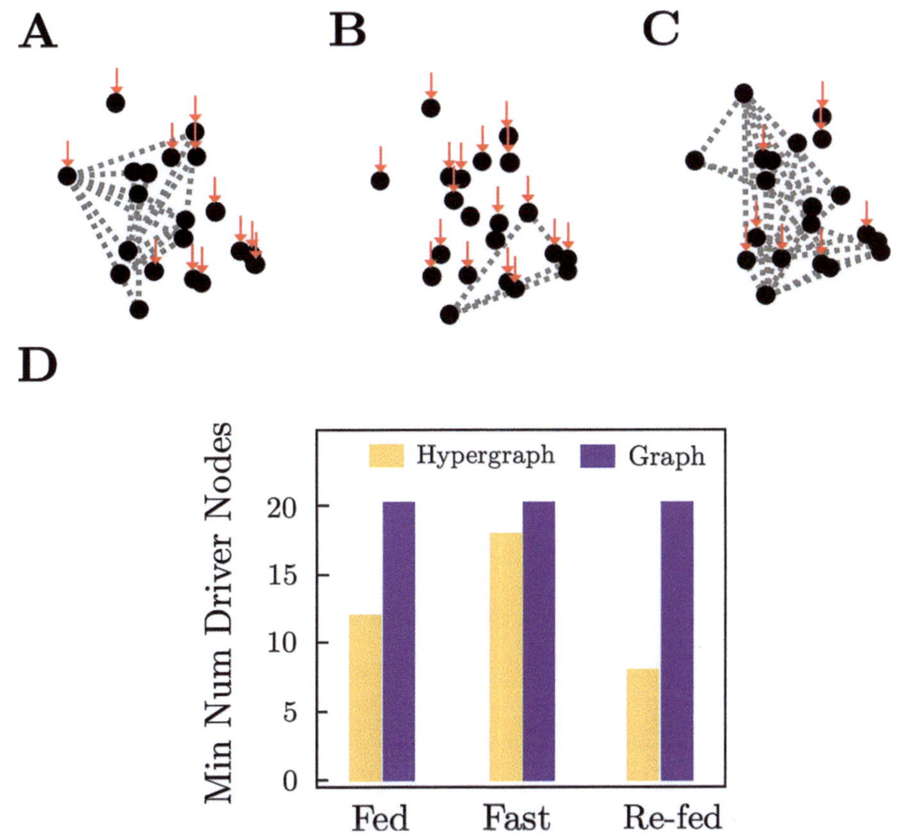

Fig. 4.4 Mouse neuron endomicroscopy dataset. **A**, **B** and **C** 3-uniform hypergraphs constructed from the neuronal activity of a mouse under the fed, fast, and re-fed phases. Each 2-simplex (i.e., a triangle) represents a hyperedge, and red arrows indicate the driver nodes. **D** Minimum number of driver nodes for 3-uniform hypergraphs under the three phases. The multi-correlation/correlation cutoff threshold is 0.95 in constructing the 3-uniform hypergraphs and the graphs. This figure was redrawn from [1] with permission

4.5.4 Chromosomal Conformation Capture Dataset

This example was adapted from [1]. The human genome can be represented by a hypergraph [59], but the technology of Hi-C can only provide pairwise genomic interactions throughout the cell cycle [60]. Given Hi-C data for a small region of chromosome 15 (100 kb bin resolution) containing two imprinted genes (*SNRPN* and *SNURF*), which are known to only express from one allele, the goal is to differentiate the differences between maternal and paternal genome architecture quantitatively throughout cell cycles. However, such differences cannot be directly observed from the Hi-C maps, as shown in Fig. 4.5A, B. Similar to the previous example, 4-uniform hypergraphs are constructed to recover the higher-order

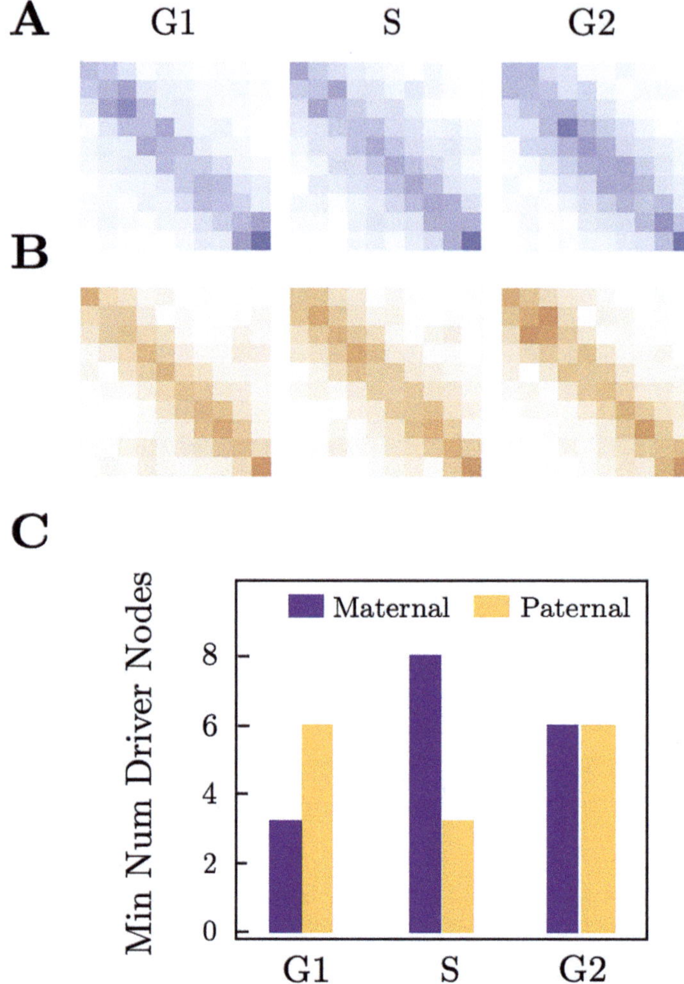

Fig. 4.5 Allele-specific Hi-C dataset. **A** and **B** Hi-C maps of a local region from the maternal and paternal Chromosome 15 during the cell cycle phases G1, S, and G2. The darker the color, the more interactions between two genomic loci. **C** Minimum numbers of driver nodes of 4-uniform hypergraphs constructed from the maternal and paternal Hi-C measurements at the G1, S, and G2 phases. The multi-correlation cutoff threshold is 0.99 in constructing the 4-uniform hypergraphs. This figure was redrawn from [1] with permission

interactions in the genome using the multi-correlation (4.20). Calculating the minimum number of driver nodes can effectively quantify the structural discrepancy between maternal and paternal genomes, as shown in Fig. 4.5C. Interestingly, the minimum number of driver nodes is equal between the maternal and paternal hypergraphs in the cell cycle phase G2, indicating some similarities in the maternal and paternal genome structures in this phase. Biologically, both genomes prepare for mitosis in the G2 phase, which may result in a small structural discrepancy.

References

1. Can, Chen, Amit Surana, Anthony M. Bloch, and Indika Rajapakse. 2021. Controllability of hypergraphs. *IEEE Transactions on Network Science and Engineering* 8 (2):1646–1657.
2. Chen, Can. 2023. Explicit solutions and stability properties of homogeneous polynomial dynamical systems. *IEEE Transactions on Automatic Control* 68 (8): 4962–4969.
3. Kruppa, Kai. 2017. Comparison of tensor decomposition methods for simulation of multilinear time-invariant systems with the mti toolbox. *IFAC-PapersOnLine* 50 (1): 5610–5615.
4. Kai, Kruppa and Gerwald Lichtenberg. 2018. Feedback linearization of multilinear time-invariant systems using tensor decomposition methods. In *SIMULTECH*, 232–243.
5. Qi, Liqun. 2005. Eigenvalues of a real supersymmetric tensor. *Journal of Symbolic Computation* 40 (6): 1302–1324.
6. Tamara, G. Kolda and Brett W. Bader. 2009. Tensor decompositions and applications. *SIAM Review* 51 (3):455–500.
7. Robeva, Elina. 2016. Orthogonal decomposition of symmetric tensors. *SIAM Journal on Matrix Analysis and Applications* 37 (1): 86–102.
8. Berge, C. 1989. *Hypergraphs, Combinatorics of Finite Sets*, 3rd ed. Amsterdam: North-Holland.
9. Eyal, Bairey, Eric D. Kelsic, and Roy Kishony. 2016. High-order species interactions shape ecosystem diversity. *Nature Communications* 7 (1):1–7.
10. Jacopo, Grilli, György Barabás, Matthew J. Michalska-Smith, and Stefano Allesina. 2017. Higher-order interactions stabilize dynamics in competitive network models. *Nature* 548 (7666):210–213.
11. Vijaysekhar, Chellaboina, Sanjay P. Bhat, Wassim M. Haddad, and Dennis S. 2009. Bernstein. Modeling and analysis of mass-action kinetics. *IEEE Control Systems Magazine* 29 (4):60–78.
12. Craciun, Gheorghe, Yangzhong Tang, and Martin Feinberg. 2006. Understanding bistability in complex enzyme-driven reaction networks. *Proceedings of the National Academy of Sciences* 103 (23): 8697–8702.
13. Donnell, Pete, and Murad Banaji. 2013. Local and global stability of equilibria for a class of chemical reaction networks. *SIAM Journal on Applied Dynamical Systems* 12 (2): 899–920.
14. Zhao, Pengcheng, Shankar Mohan, and Ram Vasudevan. 2019. Optimal control of polynomial hybrid systems via convex relaxations. *IEEE Transactions on Automatic Control* 65 (5): 2062–2077.
15. Samardzija, Nikola. 1983. Stability properties of autonomous homogeneous polynomial differential systems. *Journal of Differential Equations* 48 (1): 60–70.
16. Amir Ali Ahmadi and Bachir El Khadir. 2019. On algebraic proofs of stability for homogeneous vector fields. *IEEE Transactions on Automatic Control* 65 (1): 325–332.
17. Cunis, Torbjørn, Jean-Philippe. Condomines, and Laurent Burlion. 2020. Local stability analysis for large polynomial spline systems. *Automatica* 113: 108773.

18. Hongmei, Xie, Jun Lin, Zhiyuan Yan, and Bruce W Suter. 2012. Linearized polynomial interpolation and its applications. *IEEE Transactions on Signal Processing* 61 (1):206–217.

19. Can, Chen. 2024. On the stability of discrete-time homogeneous polynomial dynamical systems. *Computational and Applied Mathematics* 43: 75.

20. Joshua, Pickard, Amit Surana, Anthony Bloch, and Indika Rajapakse. 2023. Observability of hypergraphs. *IEEE Conference on Decision and Control*.

21. Goriely, Alain, and Craig Hyde. 1998. Finite-time blow-up in dynamical systems. *Physics Letters A* 250 (4–6): 311–318.

22. Peter, B. Stacey, Mark, L. Taper, and Veronica A. Johnson. 1997. Migration within metapopulations: the impact upon local population dynamics. In *Metapopulation biology*, 267–291. Elsevier.

23. Engen, Steinar, Russell Lande, and Bernt-Erik. Sæther. 2002. Migration and spatiotemporal variation in population dynamics in a heterogeneous environment. *Ecology* 83 (2): 570–579.

24. Qin, S. Joe, and Thomas A. Badgwell. 1997. An overview of industrial model predictive control technology. In *AIche symposium series*, vol. 93, 232–256. New York, NY: American Institute of Chemical Engineers, 1971–c2002.

25. Frank, Allgower, Rolf Findeisen, Zoltan K Nagy, et al. 2004. Nonlinear model predictive control: From theory to application. *Journal-Chinese Institute of Chemical Engineers* 35 (3):299–316.

26. Chelo, Ferreira, José L. López, and Ester P.érez Sinusía. 2006. The gauss hypergeometric function f (a, b; c; z) for large c. *Journal of Computational and Applied Mathematics* 197 (2):568–577.

27. Jan, Chyan-Deng., and Cheng-lung Chen. 2012. Use of the gaussian hypergeometric function to solve the equation of gradually-varied flow. *Journal of Hydrology* 456: 139–145.

28. Erich, Kamke. 2013. *Differentialgleichungen lösungsmethoden und lösungen*. Springer.

29. Aleksandr, Ivanovich Egorov. 2007. *Riccati equations*. Number 5. Pensoft Publishers.

30. George, Gasper and Mizan Rahman. 2004. *Basic hypergeometric series*, vol. 96. Cambridge University Press.

31. Nico, Vervliet, Otto Debals, and Lieven De Lathauwer. 2016. Tensorlab 3.0—numerical optimization strategies for large-scale constrained and coupled matrix/tensor factorization. In *2016 50th Asilomar Conference on Signals, Systems and Computers*, 1733–1738. IEEE.

32. Jurdjevic, V., and I. Kupka. 1985. Polynomial control systems. *Math. Ann.* 272: 361–368.

33. Pavol, Brunovsky. 1976. Local controllability of odd systems. *Banach Center Publications* 1 :39–458.

34. Baillieul, John. 1981. Controllability and observability of polynomial dynamical systems. *Nonlinear Analysis: Theory, Methods & Applications* 5 (5): 543–552.

35. Aeyels, Dirk. 1984. Local and global controllability for nonlinear systems. *Systems & Control Letters* 5 (1): 19–26.

36. Melody, James, Tamer Basar, and Francesco Bullo. 2003. On nonlinear controllability of homogeneous systems linear in control. *IEEE Transactions on Automatic Control* 48 (1): 139–143.

37. Anthony, M. Bloch. 2003. An introduction to aspects of geometric control theory. In *Nonholonomic mechanics and control*, 199–224. Springer.

38. Lin, Ching-Tai. 1974. Structural controllability. *IEEE Transactions on Automatic Control* 19 (3): 201–208.

39. Herbert, G. Tanner. 2004. On the controllability of nearest neighbor interconnections. In *2004 43rd IEEE conference on decision and control (CDC) (IEEE Cat. No. 04CH37601)*, vol. 3, 2467–2472. IEEE.

40. Liu, Yang-Yu., Jean-Jacques. Slotine, and Albert-László. Barabási. 2011. Controllability of complex networks. *Nature* 473 (7346): 167–173.

41. Yuan, Zhengzhong, Chen Zhao, Zengru Di, Wen-Xu. Wang, and Ying-Cheng. Lai. 2013. Exact controllability of complex networks. *Nature Communications* 4 (1): 1–9.

42. Commault, Christian. 2019. Structural controllability of networks with dynamical structured nodes. *IEEE Transactions on Automatic Control* 65 (6): 2736–2742.

43. Rahmani, Amirreza, Meng Ji, Mehran Mesbahi, and Magnus Egerstedt. 2009. Controllability of multi-agent systems from a graph-theoretic perspective. *SIAM Journal on Control and Optimization* 48 (1): 162–186.

44. Liu, Yang-Yu., and Albert-László. Barabási. 2016. Control principles of complex systems. *Reviews of Modern Physics* 88 (3): 035006.

45. Chen, Can, Chen Liao, and Yang-Yu. Liu. 2023. Teasing out missing reactions in genome-scale metabolic networks through deep learning. *Nature Communications* 14 (2375): 1–11.

46. Naganand, Yadati, Vikram Nitin, Madhav Nimishakavi, Prateek Yadav, Anand Louis, and Partha Talukdar. 2020. Nhp: Neural hypergraph link prediction. In *Proceedings of the 29th ACM international conference on information & knowledge management*, 1705–1714.

47. Yi, Han, Bin Zhou, Jian Pei, and Yan Jia. 2009. Understanding importance of collaborations in co-authorship networks: a supportiveness analysis approach. In *Proceedings of the 2009 SIAM international conference on data mining*, 1112–1123. SIAM.

48. Naganand, Yadati, Madhav Nimishakavi, Prateek Yadav, Vikram Nitin, Anand Louis, and Partha Talukdar. 2019. Hypergcn: a new method for training graph convolutional networks on hypergraphs. *Advances in Neural Information Processing Systems*, 32.

49. Geon, Lee and Kijung Shin. 2021. Thyme+: temporal hypergraph motifs and fast algorithms for exact counting. In *2021 IEEE international conference on data mining (ICDM)*, 310–319. IEEE.

50. Can, Chen and Yang-Yu Liu. 2023. A survey on hyperlink prediction. *IEEE Transactions on Neural Networks and Learning Systems*.

51. Hwang, TaeHyun, Ze Tian, Rui Kuangy, and Jean-Pierre Kocher. 2008. Learning on weighted hypergraphs to integrate protein interactions and gene expressions for cancer outcome prediction. In *2008 Eighth IEEE international conference on data mining*, 293–302. IEEE.

52. Kevin, A Murgas, Emil Saucan, and Romeil Sandhu. 2022. Hypergraph geometry reflects higher-order dynamics in protein interaction networks. *Scientific Reports* 12 (1):1–12.

53. Cooper, Joshua, and Aaron Dutle. 2012. Spectra of uniform hypergraphs. *Linear Algebra and its Applications* 436 (9): 3268–3292.

54. Hermann, Robert, and Arthur Krener. 1977. Nonlinear controllability and observability. *IEEE Transactions on Automatic Control* 22 (5): 728–740.

55. Eduardo, D. 1984. Sontag. A concept of local observability. *Systems & Control Letters* 5(1):41–47.

56. Alexandre, Sedoglavic. 2001. A probabilistic algorithm to test local algebraic observability in polynomial time. In *Proceedings of the 2001 international symposium on symbolic and algebraic computation*, 309–317.

57. Patrick, Sweeney, Can Chen, Indika Rajapakse, and Roger D Cone. 2021. Network dynamics of hypothalamic feeding neurons. *Proceedings of the National Academy of Sciences* 118 (14):e2011140118.

58. Jianji, Wang and Nanning Zheng. 2014. Measures of correlation for multiple variables. *arXiv preprint*.

59. Gabrielle, A Dotson, Can Chen, Stephen Lindsly, Anthony Cicalo, Sam Dilworth, Charles Ryan, Sivakumar Jeyarajan, Walter Meixner, Cooper Stansbury, Joshua Pickard, et al. (2022). Deciphering multi-way interactions in the human genome. *Nature Communications* 13 (1):5498.

60. Stephen, Lindsly, Wenlong Jia, Haiming Chen, Sijia Liu, Scott Ronquist, Can Chen, Xingzhao Wen, Cooper Stansbury, Gabrielle A Dotson, Charles Ryan, et al. 2021. Functional organization of the maternal and paternal human 4d nucleome. *IScience* 24 (12):103452.

Contracted Product-Based Dynamical Systems 5

Abstract

In 2017, Kruppa [*IFAC-PapersOnLine*] introduced the contracted product-based dynamical system (CPDS) representation, which expresses the system evolution as the contracted product between a dynamic tensor and a monomial state tensor. Unlike TVPDSs, this representation belongs to the class of polynomial dynamical systems. The key strength of the CPDS representation lies in the ability to perform efficient computations using tensor decompositions, including CPD, TD, TTD, and HTD. Notably, it enables the computation of analytically exact arithmetic operations such as derivatives and multiplications, which holds particular significance in developing controller algorithms like feedback linearization or model predictive control. Furthermore, CPDSs can be efficiently linearized without the need to recompute the full tensor representation.

5.1 Overview

While the TVPDS (4.1) encompasses the entire class of homogeneous polynomial dynamical systems, another tensor-based dynamical system, known as the contracted product-based dynamical system (CPDS), was introduced in [1–4] to explicitly represent polynomial dynamical systems. Unlike TVPDSs, a CPDS models system evolution using the contracted product between a two-dimensional dynamic tensor and a monomial state tensor. This monomial tensor can be visualized as a two-dimensional hypercube, where each vertex corresponds to a monomial term, see Fig. 5.1. Significantly, the CPDS representation offers improved memory efficiency compared to TVPDSs for representing polynomial dynamical systems, where the latter representation requires introducing additional dummy variables.

The continuous-time CPDS with control is defined as

$$\dot{\mathbf{x}}(t) = \langle \mathsf{A} \mid \mathsf{X}(t) \circ \mathsf{U}(t) \rangle, \tag{5.1}$$

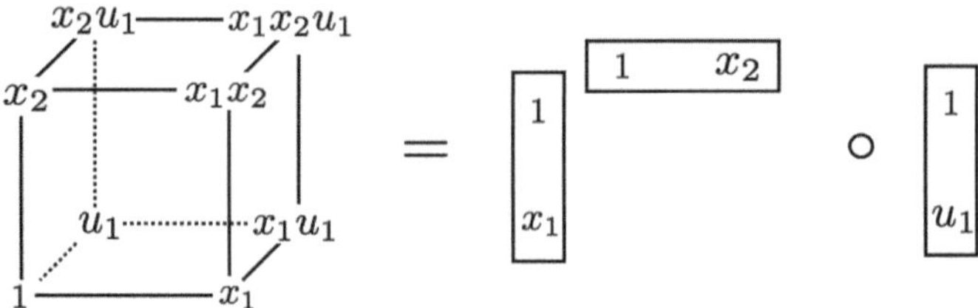

Fig. 5.1 An example of a monomial tensor $X(t) \circ U(t)$ with $n = 2$ and $s = 1$

where $\mathsf{A} \in \mathbb{R}^{(2 \times 2 \times \overset{n+s}{\cdots} \times 2) \times n}$ is a $(n + s + 1)$th-order tensor, and

$$
X(t) = \begin{bmatrix} 1 \\ x_n(t) \end{bmatrix} \circ \begin{bmatrix} 1 \\ x_{n-1}(t) \end{bmatrix} \circ \cdots \circ \begin{bmatrix} 1 \\ x_1(t) \end{bmatrix} \in \mathbb{R}^{2 \times 2 \times \overset{n}{\cdots} \times 2},
$$

$$
U(t) = \begin{bmatrix} 1 \\ u_s(t) \end{bmatrix} \circ \begin{bmatrix} 1 \\ u_{s-1}(t) \end{bmatrix} \circ \cdots \circ \begin{bmatrix} 1 \\ u_1(t) \end{bmatrix} \in \mathbb{R}^{2 \times 2 \times \overset{s}{\cdots} \times 2}.
$$

The contracted product $\langle \cdot \mid \cdot \rangle$ can be understood as the inner product between the first $n + s$ modes of the dynamic tensor A and $X(t) \circ U(t)$ for n times to produce an n-dimensional system. Similar to TVPDSs, the CPDS representation also has an unfolded form, i.e.,

$$
\dot{\mathbf{x}}(t) = \mathbf{A}_{(n+s+1)}\big(\tilde{\mathbf{u}}(t) \otimes \tilde{\mathbf{x}}(t)\big), \tag{5.2}
$$

where $\mathbf{A}_{(n+s+1)} \in \mathbb{R}^{n \times 2^{n+s}}$ is the $(n + s + 1)$-mode matricization, and

$$
\tilde{\mathbf{x}}(t) = \begin{bmatrix} 1 \\ x_1(t) \end{bmatrix} \otimes \begin{bmatrix} 1 \\ x_2(t) \end{bmatrix} \otimes \cdots \otimes \begin{bmatrix} 1 \\ x_n(t) \end{bmatrix} \in \mathbb{R}^{2^n},
$$

$$
\tilde{\mathbf{u}}(t) = \begin{bmatrix} 1 \\ u_1(t) \end{bmatrix} \otimes \begin{bmatrix} 1 \\ u_2(t) \end{bmatrix} \otimes \cdots \otimes \begin{bmatrix} 1 \\ u_s(t) \end{bmatrix} \in \mathbb{R}^{2^s}.
$$

Polynomial dynamical systems offer a broader range of applications, spanning from robotics [5, 6], power systems [7, 8], aerospace [9, 10], ecological systems [11, 12], and biomedical engineering [13, 14]. Notably, many of these systems involve an exceptionally high number of states. Employing the full tensor representation of the CPDS (5.1) therefore would result in a significant memory demand due to the exponential increase in the number of parameters in the dynamic tensor A, i.e., $2^{n+s}n$. To address this challenge, tensor decomposition techniques such as CPD, TD, TTD, and HTD can be effectively leveraged to significantly reduce the complexity and memory requirements. Note that homogenizing polynomial dynamical systems to TVPDSs incurs even higher memory demands.

This chapter explores the applications of four tensor decomposition techniques, namely CPD, TD, TTD, and HTD, for CPDSs. By decomposing the dynamic tensor, these techniques facilitate more efficient computations with reduced memory requirements. Additionally, a linearization method for CPDSs based on decomposition factors is presented. Furthermore, feedback linearization for CPDSs using tensor decompositions is discussed. The content of this chapter is primarily based on the work of [1, 2].

5.2   Decomposed System Representations

This section discusses four different tensor decomposition forms of the CPDS (5.1). These forms have the potential to greatly decrease the computational and memory complexity.

5.2.1   CPD

The CPD form of the CPDS (5.1) can be illustrated using the following simple example:

$$\begin{cases} \dot{x}_1(t) = a_1 + a_2 x_1(t) + a_3 x_1(t) u_1(t) \\ \dot{x}_2(t) = b_1 x_2(t) + b_2 x_1(t) x_2(t) + b_3 x_1(t) x_2(t) u_1(t) \end{cases} ,$$

where the CPD of the dynamic tensor $\mathbf{A}$ can be represented as

$$\mathbf{A}_{x_1} = \begin{bmatrix} 1 & 0 & 0 & 1 & 0 & 0 \\ 0 & 1 & 1 & 0 & 1 & 1 \end{bmatrix}, \quad \mathbf{A}_{x_2} = \begin{bmatrix} 1 & 1 & 1 & 0 & 0 & 0 \\ 0 & 0 & 0 & 1 & 1 & 1 \end{bmatrix},$$

$$\mathbf{A}_{u_1} = \begin{bmatrix} 1 & 1 & 0 & 1 & 1 & 0 \\ 0 & 0 & 1 & 0 & 0 & 1 \end{bmatrix}, \quad \mathbf{A}_f = \begin{bmatrix} 1 & 1 & 1 & 0 & 0 & 0 \\ 0 & 0 & 0 & 1 & 1 & 1 \end{bmatrix},$$

$$\boldsymbol{\lambda} = \begin{bmatrix} a_1 & a_2 & a_3 & b_1 & b_2 & b_3 \end{bmatrix}.$$

The factor matrices can be constructed column-wise using coefficients a_j or b_j that correspond to each column. If the state or control inputs in the polynomial dynamical system is multiplied by the coefficient, the column of the factor matrices $\mathbf{A}_{x_j}$ or $\mathbf{A}_{u_j}$ is set to $\begin{bmatrix} 0 & 1 \end{bmatrix}^\top$. Otherwise, it is set to $\begin{bmatrix} 1 & 0 \end{bmatrix}^\top$. The last factor matrix $\mathbf{A}_f$ shows to which state equations the coefficient belongs. Therefore, every CPDS can be represented in the CPD form.

Proposition 5.1 ([1]) *The CPDS (5.1) can be represented in the CPD form as*

$$\dot{x}(t) = A_f \left(\lambda \odot \left(A_{u_s}^\top \begin{bmatrix} 1 \\ u_s(t) \end{bmatrix} \right) \odot \cdots \odot \left(A_{u_1}^\top \begin{bmatrix} 1 \\ u_1(t) \end{bmatrix} \right) \right.$$

$$\left. \odot \left(A_{x_n}^\top \begin{bmatrix} 1 \\ x_n(t) \end{bmatrix} \right) \odot \cdots \odot \left(A_{x_1}^\top \begin{bmatrix} 1 \\ x_1(t) \end{bmatrix} \right) \right). \tag{5.3}$$

More importantly, the CPD form of CPDSs can help to construct systems representations with TD, TTD, and HTD.

5.2.2　TD

As mentioned previously, every CPD can be rewritten into the TD representation. Therefore, it is only required to translate the weight vector λ from the CPD form to a diagonal tensor for obtaining the TD form of the CPDS (5.1).

Proposition 5.2 ([1]) *The CPDS (5.1) can be represented in the TD form as*

$$
\begin{aligned}
\dot{x}(t) = A_f \Bigg(S \times_1 \left(A_{u_s}^\top \begin{bmatrix} 1 \\ u_s(t) \end{bmatrix} \right) \times_2 \cdots \times_s \left(A_{u_1}^\top \begin{bmatrix} 1 \\ u_1(t) \end{bmatrix} \right) \\
\times_{s+1} \left(A_{x_n}^\top \begin{bmatrix} 1 \\ x_n(t) \end{bmatrix} \right) \times_{s+2} \cdots \times_{s+n} \left(A_{x_1}^\top \begin{bmatrix} 1 \\ x_1(t) \end{bmatrix} \right) \Bigg),
\end{aligned}
\tag{5.4}
$$

where S is a diagonal tensor containing the weights λ from the CPD form along its diagonal.

The full TD form can be demanding on memory resources. As a result, truncation techniques may be employed to decrease the size of the core tensor S.

5.2.3　TTD

The TTD form of the CPDS (5.1) can also be translated from its CPD form by setting the core tensors as follows:

$$
\begin{aligned}
(\mathbf{A}_{u_s})_{j_1:} &= (\mathbf{A}_{u_s})_{j_1:}, \\
(\mathbf{A}_{u_i})_{:j_{s-i+1}:} &= \mathrm{diag}\!\left((\mathbf{A}_{u_i})_{j_{s-i+1}:} \right) && \text{for } i = 1, 2, \ldots, s-1, \\
(\mathbf{A}_{x_i})_{:j_{n+s-i+1}:} &= \mathrm{diag}\!\left((\mathbf{A}_{s_i})_{j_{n+s-i+1}:} \right) && \text{for } i = 1, 2, \ldots, n, \\
(\mathbf{A}_f)_{:j_{n+s+1}} &= \lambda \odot (\mathbf{A}_f)_{j_{n+s+1}:}^\top.
\end{aligned}
$$

Proposition 5.3 ([1]) *The CPDS (5.1) can be represented in the TTD form as*

$$
\begin{aligned}
\dot{x}(t) = \left(A_{u_s} \times_2 \begin{bmatrix} 1 \\ u_s(t) \end{bmatrix} \right) \cdots \left(A_{u_1} \times_2 \begin{bmatrix} 1 \\ u_1(t) \end{bmatrix} \right) \\
\left(A_{x_n} \times_2 \begin{bmatrix} 1 \\ x_n(t) \end{bmatrix} \right) \cdots \left(A_{x_1} \times_2 \begin{bmatrix} 1 \\ x_1(t) \end{bmatrix} \right) A_f.
\end{aligned}
\tag{5.5}
$$

Much like in the case of TD, the direct transformation from CPD to TTD can be memory-intensive. Therefore, truncation techniques that employ singular value decomposition can be utilized to attain lower TT-ranks.

5.2.4 HTD

The direct conversion from the CPD form of the CPDS (5.1) to its HTD form can be achieved by constructing a tree for the dynamic tensor A. The leaf node matrices are equal to the factor matrices $\mathbf{A}_{u_i}$, $\mathbf{A}_{x_i}$, and $\mathbf{A}_f$ in the CPD form. The transfer function $\mathbf{G}_{u_s \cdots u_1 x_n \cdots x_1 f}$ of the top node is a diagonal matrix containing the CPD weights λ along its diagonal, and other transfer functions are set to

$$\mathbf{G} = 1_{(2,\{1,2,3\})},$$

where $1 \in \mathbb{R}^{r \times r \times r}$ is a diagonal tensor that contains ones along its diagonal. Here r is the length of the weight vector λ.

Proposition 5.4 ([1]) *The CPDS (5.1) can be represented in the THD form as*

$$\dot{x}(t) = A \times_1 \begin{bmatrix} 1 \\ u_s(t) \end{bmatrix} \times_2 \cdots \times_s \begin{bmatrix} 1 \\ u_1(t) \end{bmatrix} \times_{s+1} \begin{bmatrix} 1 \\ x_n(t) \end{bmatrix} \times_{s+2} \cdots \times_{s+n} \begin{bmatrix} 1 \\ x_1(t) \end{bmatrix}, \qquad (5.6)$$

where the p-mode products can be computed by replacing the leaf nodes A_{u_i} and A_{x_i} in the tree of A with $\begin{bmatrix} 1 & u_i(t) \end{bmatrix} A_{u_i}$ and $\begin{bmatrix} 1 & x_i(t) \end{bmatrix} A_{x_i}$.

Again, the conversion from the CPD form to the HTD form of the CPDS (5.1) can also be inefficient in terms of memory usage. However, truncation techniques can be utilized by leveraging the SVD of the matricizations of the original tensor, which can help to optimize the process.

5.3 Linearization

Linearization is a well-established technique for analyzing nonlinear dynamical systems [15–17]. Tensor-based differentiation provides an efficient approach for linearizing CPDSs.

Lemma 5.1 ([1]) *Given an n-dimensional polynomial equation represented as*

$$h(x) = \langle A, X \rangle,$$

where $A \in \mathbb{R}^{2 \times 2 \times \overset{n}{\cdots} \times 2}$, and

$$X = \begin{bmatrix} 1 \\ x_n \end{bmatrix} \circ \begin{bmatrix} 1 \\ x_{n-1} \end{bmatrix} \circ \cdots \circ \begin{bmatrix} 1 \\ x_1 \end{bmatrix} \in \mathbb{R}^{2 \times 2 \times \overset{n}{\cdots} \times 2},$$

the partial derivative of $h(x)$ with respect to x_j can be computed as

$$\frac{\partial}{\partial x_j} h(x) = \langle A_{x_j}, X \rangle, \tag{5.7}$$

where $A_{x_j} = A \times_{n-j+1} \begin{bmatrix} 0 & 1 \\ 0 & 0 \end{bmatrix}$.

Therefore, the linearization of the CPDS (5.1) can be readily obtained using the above lemma.

Proposition 5.5 ([1]) *The linearized system matrices for the CPDS (5.1) can be analytically derived as*

$$A = \langle A_{lin} \mid X_e \circ U_e \rangle \in \mathbb{R}^{n \times n},$$
$$B = \langle B_{lin} \mid X_e \circ U_e \rangle \in \mathbb{R}^{n \times s},$$

where $A_{lin} \in \mathbb{R}^{(2 \times 2 \times \overset{n}{\cdots} \times 2) \times n \times n}$ *and* $B_{lin} \in \mathbb{R}^{(2 \times 2 \times \overset{n}{\cdots} \times 2) \times n \times s}$ *are* $(n+2)$*th-order tensors such that*

$$(A_{lin})_{:::\cdots:j} = A_{x_j} \text{ for } j = 1, 2, \ldots, n,$$
$$(B_{lin})_{:::\cdots:j} = A_{u_j} \text{ for } j = 1, 2, \ldots, s.$$

The linearized system matrices $\mathbf{A}$ and $\mathbf{B}$ can be directly computed from the dynamic tensor A using the contracted and p-mode tensor matrix products. This computation is particularly efficient if the CPDS is provided in the CPD, TD, TTD, or HTD form without converting to the full tensors [1].

5.4   Feedback Linearization

Feedback linearization is a control design technique that can convert an input affine nonlinear system into a linear system with a change of variables and a feedback control law [18–21]. This technique involves applying a nonlinear transformation that maps the original nonlinear system dynamics to a linear system dynamics and then designing a feedback control based on the transformed system. Consider a nonlinear input affine single-input single-output (SISO) system

$$\begin{cases} \dot{\mathbf{x}}(t) = \mathbf{f}(\mathbf{x}(t)) + \mathbf{g}(\mathbf{x}(t)) u(t) \\ y = c(\mathbf{x}(t)) \end{cases}, \tag{5.8}$$

where $\mathbf{f}$ and $\mathbf{g}$ are nonlinear vector fields, and c is an output scalar function. Suppose that $\mathbf{f}$, $\mathbf{g}$, and c are sufficiently smooth. The control law with a reference input r based on feedback linearization is given by

$$u = \frac{-\sum_{j=0}^{q} \alpha_j L_{\mathbf{f}}^{j} c(\mathbf{x}) + \alpha_0 r}{L_{\mathbf{g}} L_{\mathbf{f}}^{q-1} c(\mathbf{x})}. \tag{5.9}$$

The closed-loop system exhibits a linear response from the reference input r to the system output y governed by the following linear differential equation:

$$\alpha_q \frac{\partial^q}{\partial t^q} y + \alpha_{q-1} \frac{\partial^{q-1}}{\partial t^{q-1}} y + \cdots + \alpha_1 \frac{\partial}{\partial t} y + \alpha_0 y = \alpha_0 r.$$

The Lie derivative of the scalar function c along a vector field $\mathbf{f}$ is defined as

$$L_{\mathbf{f}} c(\mathbf{x}) = \sum_{i=1}^{n} f_i(\mathbf{x}) \frac{\partial c(\mathbf{x})}{\partial x_i} = \langle \mathbf{f}(\mathbf{x}), \nabla c(\mathbf{x}) \rangle,$$

and $L_{\mathbf{f}}^{j} c(\mathbf{x})$ denotes the jth Lie derivative.

5.4.1 Higher-Order Polynomials

Computing the Lie derivative involves higher-order terms, making it necessary to introduce higher-order polynomials. Define higher-order polynomials in the following form:

$$h(\mathbf{x}) = \langle \mathsf{H}, \mathsf{X}^l \rangle, \tag{5.10}$$

where $\mathsf{H} \in \mathbb{R}^{(2 \times 2 \times \overset{nl}{\cdots} \times 2) \times n}$ is a $(nl + 1)$th-order tensor, and

$$\mathsf{X}^l = \left(\begin{bmatrix} 1 \\ x_n \end{bmatrix} \circ \begin{bmatrix} 1 \\ x_{n-1} \end{bmatrix} \circ \cdots \circ \begin{bmatrix} 1 \\ x_1 \end{bmatrix} \right) \circ \overset{l}{\cdots} \circ \left(\begin{bmatrix} 1 \\ x_n \end{bmatrix} \circ \begin{bmatrix} 1 \\ x_{n-1} \end{bmatrix} \circ \cdots \circ \begin{bmatrix} 1 \\ x_1 \end{bmatrix} \right) \in \mathbb{R}^{2 \times 2 \times \overset{nl}{\cdots} \times 2}.$$

In the higher-order form (5.10), the maximal order of each x_j in the monomials with exponents up to l. Many arithmetic operations including multiplication, differentiation, and Lie derivative can be computed in the tensor form [2].

Proposition 5.6 ([2]) *Given two n-dimensional polynomials of the forms*

$$h_1(\boldsymbol{x}) = \langle \mathsf{H}_1, \mathsf{X}^{l_1} \rangle \ \text{and} \ h_2(\boldsymbol{x}) = \langle \mathsf{H}_2, \mathsf{X}^{l_2} \rangle,$$

the product of the two polynomial can be computed as

$$h_1(\boldsymbol{x}) h_2(\boldsymbol{x}) = \langle \mathsf{H}_1 \circ \mathsf{H}_2, \mathsf{X}^{(l_1 + l_2)} \rangle. \tag{5.11}$$

The following result generalizes Lemma 5.1. Let $\mathbf{P} = \begin{bmatrix} 0 & 1 \\ 0 & 0 \end{bmatrix}$.

Proposition 5.7 ([2]) *Given an n-dimensional polynomial of the form*

$$h(\boldsymbol{x}) = \langle \mathsf{H}, \mathsf{X}^l \rangle,$$

the partial derivative of h with respect to x_j can be computed as

$$\frac{\partial}{\partial x_j} h(x) = \langle H_j, X^l \rangle, \tag{5.12}$$

where $H_j = \sum_{i=1}^{l} H \times_{in-j+1} P$.

Proposition 5.8 ([2]) *Given an n-dimensional polynomial system and a polynomial of the forms*

$$h(x) = \langle H \mid X^l \rangle \text{ and } c(x) = \langle G, X^l \rangle,$$

respectively, the Lie derivative of c along h can be computed as

$$L_h^j = \langle L_j \mid X^{l(j+1)} \rangle, \tag{5.13}$$

where

$$L_j = \begin{cases} G & \text{for } j = 0 \\ \sum_{i=1}^{n} H_{::\cdots:i} \circ \left(\sum_{k=1}^{l} G \times_{kn-i+1} P \right) & \text{for } j = 1 \\ \sum_{i=1}^{n} H_{::\cdots:i} \circ \left(\sum_{k=1}^{jl} L_{j-1} \times_{kn-i+1} P \right) & \text{otherwise} \end{cases}.$$

The proofs of the above propositions can be found in [2].

5.4.2 CPDS Feedback Linearization

Consider the following input affine SISO CPDS:

$$\begin{cases} \dot{\mathbf{x}}(t) = \langle A \mid X(t) \rangle + \langle B \mid X(t) \rangle u(t) \\ y(t) = \langle C, X(t) \rangle \end{cases}, \tag{5.14}$$

where $A \in \mathbb{R}^{(2 \times 2 \times \overset{n}{\cdots} \times 2) \times n}$, $B \in \mathbb{R}^{(2 \times 2 \times \overset{n}{\cdots} \times 2) \times n}$, and $C \in \mathbb{R}^{2 \times 2 \times \overset{n}{\cdots} \times 2}$. Assume that the CPDS (5.14) is controllable, observable, and feedback linearizable with well-defined relative degree q.

Proposition 5.9 ([2]) *The feedback linearizing controller for the input affine SISO CPDS (5.14) can be computed as*

$$u(t) = \frac{-\langle \sum_{j=0}^{q} \alpha_j L_j, X^{q+1}(t) \rangle + \alpha_0 r}{\langle K_{q-1}, X^{q+1}(t) \rangle}, \tag{5.15}$$

where

$$
L_j = \begin{cases}
C & \text{for } j = 0 \\
\sum_{i=1}^{n} A_{::\cdots:i} \circ (\sum_{k=1}^{l} C \times_{kn-i+1} P) & \text{for } j = 1, \\
\sum_{i=1}^{n} A_{::\cdots:i} \circ (\sum_{k=1}^{jl} L_{j-1} \times_{kn-i+1} P) & \text{otherwise}
\end{cases}
$$

and

$$
K_j = \sum_{i=1}^{n} B_{::\cdots:i} \circ \left(\sum_{k=1}^{j+1} L_j \times_{kn-i+1} P \right).
$$

Proof The result follows immediately from (5.9) and Proposition 5.8. $\square$

Many operations, such as summation, multiplication, outer product, and p-mode product, can be efficiently achieved using CPD, TD, TTD, and HTD, as mentioned earlier. Therefore, if CPDSs are provided in the CPD, TD, TTD, or HTD form, as described in the last section, it can significantly increase the efficiency of designing feedback linearizing controllers for large-scale systems [2]. It's worth noting that this approach can be generalized to multiple-input multiple-output (MIMO) polynomial dynamical systems as well.

5.5 Applications

5.5.1 HVAC Systems

This example, adapted from [1], considers a complex model of a heating, ventilation, and air conditioning (HVAC) system introduced in [22]. The model is created based on heat and mass balances and can be represented by a CPDS of the form (5.1) with 17 states and 10 inputs. Hence, the full tensor representation consists of $17 \times 2^{27} = 2.3 \times 10^9$ parameters. The four tensor decomposition forms (CPD, TD, TTD, and HTD) are applied to represent this CPDS. The memory requirements for the four forms of the CPDS are presented in Table 5.1. First, all four forms significantly reduce the total number of parameters in the dynamic tensor. In particular, the CPD form achieves the lowest memory demand, even though it cannot be reduced further through truncation. On the other hand, the TTD and HTD forms can be truncated by setting a desired approximation threshold without significant loss in the dynamical system behavior. Therefore, the tensor decomposition forms of the CPDS can allow for a highly efficient simulation of the system and control design (e.g., feedback linearization).

5.5.2 Synthetic Data: Feedback Linearization

This example, adapted from [2], examines an SISO input affine polynomial dynamical system (i.e., a CPDS) given by

Table 5.1 Memory requirement comparisons for the four tensor decomposition forms of the HVAC CPDS. This table was adapted from [1] with permission

	Full	CPD	TD	TTD	HTD
Exact	2.3×10^9	3834	85482	152658	4100814
Truncated	2.3×10^9	3834	85482	9779	7634

$$\begin{cases} \begin{bmatrix} \dot{x}_1(t) \\ \dot{x}_2(t) \end{bmatrix} = \begin{bmatrix} 2x_2(t) \\ -x_1(t) + 0.2x_1(t)x_2(t) \end{bmatrix} + \begin{bmatrix} 0 \\ u(t) \end{bmatrix} \\ y = 1 + 2x_1(t) \end{cases}.$$

The system can be represented in the CPD form with factor matrices

$$A = \left\{ \begin{bmatrix} 0 & 1 & 0 \\ 1 & 0 & 1 \end{bmatrix}, \begin{bmatrix} 1 & 0 & 0 \\ 0 & 1 & 1 \end{bmatrix}, \begin{bmatrix} 1 & 0 & 0 \\ 0 & 1 & 1 \end{bmatrix}, \begin{bmatrix} 2 \\ -1 \\ 0.2 \end{bmatrix} \right\},$$

$$B = \left\{ \begin{bmatrix} 1 \\ 0 \end{bmatrix}, \begin{bmatrix} 1 \\ 0 \end{bmatrix}, \begin{bmatrix} 0 \\ 1 \end{bmatrix}, 1 \right\},$$

$$C = \left\{ \begin{bmatrix} 1 & 1 \\ 0 & 0 \end{bmatrix}, \begin{bmatrix} 0 & 1 \\ 1 & 0 \end{bmatrix}, \begin{bmatrix} 2 \\ 1 \end{bmatrix} \right\}.$$

The Lie derivative can be computed in the CPD form using tensor algebra, i.e.,

$$L_1 = \left\{ \begin{bmatrix} 0 \\ 1 \end{bmatrix}, \begin{bmatrix} 1 \\ 0 \end{bmatrix}, \begin{bmatrix} 1 \\ 0 \end{bmatrix}, \begin{bmatrix} 1 \\ 0 \end{bmatrix}, 4 \right\},$$

$$L_2 = \left\{ \begin{bmatrix} 1 & 0 \\ 0 & 1 \end{bmatrix}, \begin{bmatrix} 0 & 0 \\ 1 & 1 \end{bmatrix}, \begin{bmatrix} 1 & 1 \\ 0 & 0 \end{bmatrix}, \begin{bmatrix} 1 & 1 \\ 0 & 0 \end{bmatrix}, \begin{bmatrix} 1 & 1 \\ 0 & 0 \end{bmatrix}, \begin{bmatrix} 1 & 1 \\ 0 & 0 \end{bmatrix}, \begin{bmatrix} -4 \\ 0.8 \end{bmatrix} \right\},$$

$$K_0 = \left\{ \begin{bmatrix} 0 \\ 0 \end{bmatrix}, \begin{bmatrix} 0 \\ 0 \end{bmatrix}, \begin{bmatrix} 0 \\ 0 \end{bmatrix}, \begin{bmatrix} 0 \\ 0 \end{bmatrix}, 0 \right\},$$

$$K_1 = \left\{ \begin{bmatrix} 1 \\ 0 \end{bmatrix}, \begin{bmatrix} 1 \\ 0 \end{bmatrix}, \begin{bmatrix} 1 \\ 0 \end{bmatrix}, \begin{bmatrix} 1 \\ 0 \end{bmatrix}, \begin{bmatrix} 1 \\ 0 \end{bmatrix}, \begin{bmatrix} 1 \\ 0 \end{bmatrix}, 4 \right\}.$$

The system has a relative degree of $q = 2$ and is both controllable and observable. The linear behavior in the closed loop therefore is given by

$$\alpha_2 \ddot{y} + \alpha_1 \dot{y} + \alpha_0 y = \alpha_0 r,$$

where the parameters $\alpha_0 = \alpha_1 = 10$ and $\alpha_2 = 1$ are chosen. Finally, by evaluating (5.15), the feedback linearizing controller is given by

$$u = \frac{10 + 16x_1 + 40x_2 + 0.8x_1 x_2 + 10r}{4}.$$

The effectiveness of this feedback linearizing controller can be verified with numerical simulations, where the system behaves as the specified second-order linear system.

References

1. Kruppa, Kai. 2017. Comparison of tensor decomposition methods for simulation of multilinear time-invariant systems with the mti toolbox. *IFAC-PapersOnLine* 50 (1): 5610–5615.
2. Kruppa, Kai and Gerwald Lichtenberg. 2018. Feedback linearization of multilinear time-invariant systems using tensor decomposition methods. In *SIMULTECH*, 232–243.
3. Pangalos, Georg, Annika Eichler, and Gerwald Lichtenberg. 2015. Hybrid multilinear modeling and applications. In *Simulation and modeling methodologies, technologies and applications: international conference, SIMULTECH 2013 Reykjavík, Iceland, 2013 Revised Selected Papers*, 71–85. Berlin: Springer.
4. Lichtenberg, Gerwald. 2011. *Hybrid tensor systems*. Ph.D. thesis
5. Richards, Spencer M., Felix Berkenkamp, and Andreas Krause. 2018. The lyapunov neural network: Adaptive stability certification for safe learning of dynamical systems. In *Conference on robot learning*, 466–476. PMLR.
6. Capco, Jose, Mohab Safey El. Din, and Josef Schicho. 2023. Positive dimensional parametric polynomial systems, connectivity queries and applications in robotics. *Journal of Symbolic Computation* 115: 320–345.
7. Zhou, Yongzhi, Wu. Hao, Gu. Chenghong, and Yonghua Song. 2016. A novel method of polynomial approximation for parametric problems in power systems. *IEEE Transactions on Power Systems* 32 (4): 3298–3307.
8. Xiao, Weidong, Magnus GJ Lind, William G Dunford, and Antoine Capel. 2006. Real-time identification of optimal operating points in photovoltaic power systems. *IEEE Transactions on industrial Electronics* 53 (4): 1017–1026.
9. Zheng, Xiaohu, Wen Yao, Yunyang Zhang, and Xiaoya Zhang. 2022. Parameterized consistency learning-based deep polynomial chaos neural network method for reliability analysis in aerospace engineering. arXiv:2203.15655.
10. Iyer, Ganesh Neelakanta, Bharadwaj Veeravalli, and Sakthi Ganesh Krishnamoorthy. 2012. On handling large-scale polynomial multiplications in compute cloud environments using divisible load paradigm. *IEEE Transactions on Aerospace and Electronic Systems* 48 (1): 820–831.
11. Sabbar, Yassine, and Driss Kiouach. 2023. New method to obtain the acute sill of an ecological model with complex polynomial perturbation. *Mathematical Methods in the Applied Sciences* 46 (2): 2455–2474.
12. Wörz-Busekros, Angelika. 1978. Global stability in ecological systems with continuous time delay. *SIAM Journal on Applied Mathematics* 35 (1): 123–134.
13. Dam, Jan S., Torben Dalgaard, Paul Erik Fabricius, and Stefan Andersson-Engels. 2000. Multiple polynomial regression method for determination of biomedical optical properties from integrating sphere measurements. *Applied Optics* 39 (7): 1202–1209.
14. Bhateja, Vikrant, Mukul Misra, Shabana Urooj, and Aimé Lay-Ekuakille. 2013. A robust polynomial filtering framework for mammographic image enhancement from biomedical sensors. *IEEE Sensors Journal* 13 (11): 4147–4156.
15. Caughey, Thomas K. 1963. Equivalent linearization techniques. *The Journal of the Acoustical Society of America* 35 (11): 1706–1711.

16. Krener, Arthur J. 1973. On the equivalence of control systems and the linearization of nonlinear systems. *SIAM Journal on Control* 11 (4): 670–676.
17. Cheng, Daizhan, Xiaoming Hu, Tielong Shen, Daizhan Cheng, Xiaoming Hu, and Tielong Shen. 2010. Linearization of nonlinear systems. In *Analysis and design of nonlinear control systems*, 279–313.
18. Krener, A. J. 1999. Feedback linearization. *Mathematical control theory*, 66–98.
19. Charlet, Bernard, Jean Lévine, and Riccardo Marino. 1989. On dynamic feedback linearization. *Systems and Control Letters* 13 (2): 143–151.
20. Khalil, Hassan K. 2015. *Nonlinear control*, vol. 406. Pearson, New York.
21. Isidori, Alberto. 1985. *Nonlinear control systems: an introduction*. Berlin: Springer.
22. Tashtoush, Bourhan, Mohammed Molhim, and Mohammed Al-Rousan. 2005. Dynamic model of an hvac system for control analysis. *Energy* 30 (10): 1729–1745.

t-Product-Based Dynamical Systems

6

Abstract

The concept of t-product-based dynamical systems (t-PDS) representation was first introduced by Hoover et al. [posted on *arXiv*] in 2021. The system evolution is governed by the t-product between a third-order dynamic tensor and a third-order state tensor. Recent advancements in tensor decomposition and the algebra of circulants facilitate a natural extension of linear systems theory to t-PDSs, encompassing concepts such as explicit solutions, stability, controllability, and observability. Furthermore, the theory of state feedback control can be extended to t-PDSs in a manner analogous to that of LDSs.

6.1 Overview

Different from the previously introduced tensor-based dynamical systems, t-product-based dynamical systems (t-PDSs) are exclusively defined over third-order tensors, with system evolution being captured by the t-product operation [1]. The continuous-time t-PDS with control inputs is defined as

$$\dot{\mathsf{X}}(t) = \mathsf{A} \star \mathsf{X}(t) + \mathsf{B} \star \mathsf{U}(t), \tag{6.1}$$

where $\mathsf{A} \in \mathbb{R}^{n \times n \times r}$ is the dynamic tensor, $\mathsf{B} \in \mathbb{R}^{n \times s \times r}$ is the control tensor, $\mathsf{X}(t) \in \mathbb{R}^{n \times h \times r}$ is the state variable, and $\mathsf{U}(t) \in \mathbb{R}^{s \times h \times r}$ is the control inputs. Similar to other tensor-based dynamical systems, the t-PDS (6.1) can be equivalently represented in an unfolded form, i.e.,

$$\dot{\mathbf{X}}(t) = \mathtt{bcirc}(\mathsf{A})\mathbf{X}(t) + \mathtt{bcirc}(\mathsf{B})\mathbf{U}(t), \tag{6.2}$$

where $\mathbf{X}(t) = \mathtt{unfold}(\mathsf{X}(t))$ and $\mathbf{U}(t) = \mathtt{unfold}(\mathsf{U}(t))$. Readers may refer to this chapter for the detailed definition of $\mathtt{bcirc}$. Consequently, linear system-theoretic properties, such

© The Author(s), under exclusive license to Springer Nature Switzerland AG 2024 97
C. Chen, *Tensor-Based Dynamical Systems*,
Synthesis Lectures on Mathematics & Statistics,
https://doi.org/10.1007/978-3-031-54505-4_6

as explicit solutions, stability, controllability, and observability, can be extended to t-PDSs based on the unfolded form.

The t-PDS framework provides a versatile approach to analyzing and controlling complex dynamical systems in a variety of fields, including physics [2, 3], engineering [4, 5], and biology [6–8]. This framework allows for the representation of system states as third-order tensors, enabling a more comprehensive and nuanced understanding of complex system behavior compared to traditional vector-based approaches. Notably, t-PDSs leverage the rich mathematical framework of tensor algebra, empowering researchers to perform sophisticated operations on multidimensional data, e.g., images. For instance, t-PDSs have potential applications in tasks such as image denoising, image compression, and image segmentation, as demonstrated by recent research [9–12].

This chapter delves into the system-theoretic properties of t-PDSs, encompassing explicit solutions, stability, controllability, and observability according to tensor decompositions and the algebra of circulants. Additionally, the chapter introduces the technique of state feedback control for t-PDSs. The content of this chapter is mainly based on the work of [1].

6.2   System-Theoretic Properties

This section explores the system-theoretic properties of t-PDSs, including explicit solutions, stability, controllability, and observability. Interestingly, the results are similar to those in LDSs.

6.2.1 Explicit Solutions

The explicit solution of the unforced t-PDS, i.e.,

$$\dot{\mathsf{X}}(t) = \mathsf{A} \star \mathsf{X}(t), \tag{6.3}$$

can be readily obtained from its unfolded form.

Proposition 6.1 ([1]) *The explicit solution of the unforced t-PDS (6.3) can be computed as*

$$X(t) = \mathtt{fold}\big(\exp\{\mathtt{bcirc}(A)t\}X(0)\big). \tag{6.4}$$

Proof The result follows immediately from the unfolded form of (6.3), which is given by $\dot{\mathbf{X}}(t) = \mathtt{bcirc}(A)\mathbf{X}(t)$. $\qquad\qquad\square$

For simplicity, the explicit solution can be denoted as $\mathsf{X}(t) = \exp\{\mathsf{A}t\}\star\mathsf{X}(0)$, which takes a similar format to that in LDSs. Therefore, the complete solution to the t-PDS (6.1) can be derived.

Proposition 6.2 ([1]) *The complete solution of the t-PDS (6.1) can be computed as*

$$X(t) = \exp\{At\} \star X(0) + \int_0^t \exp\{A(t - \tau)\} \star B \star U(\tau)d\tau. \tag{6.5}$$

Similarly, the complete solution can be computed in the unfolded form, i.e.,

$$\mathbf{X}(t) = \exp\{\texttt{bcirc(A)}t\}\mathbf{X}(0) + \int_0^t \left(\exp\{\texttt{bcirc(A)}(t - \tau)\}\texttt{unfold(B)}\right)\mathbf{U}(t)d\tau.$$

6.2.2   Stability

The stability for t-PDSs can be defined similarly as for LDSs. Analogous to the linear stability, the stability properties of the unforced t-PDS (6.3) can be determined by the t-eigenvalue decomposition of the dynamic tensor A.

Definition 6.1   The equilibrium point $X_e = 0 \in \mathbb{R}^{n \times h \times r}$ of the t-PDS (6.3) is called:

- stable if $\|X(t)\| \leq \gamma\|X(0)\|$ for some $\gamma > 0$;
- asymptotically stable if $\lim_{t \to \infty} \|X(t)\| = 0$;
- unstable if it is not stable.

Proposition 6.3   *Given the t-PDS (6.3), the equilibrium point $X_e = 0$ is:*

- *stable if and only if $\lambda_j \leq 0$ for all $j = 1, 2, \ldots, nr$;*
- *stable if and only if $\lambda_j < 0$ for all $j = 1, 2, \ldots, nr$;*
- *unstable if and only if $\lambda_j > 0$ for some $j = 1, 2, \ldots, nr$,*

where λ_j are the eigenvalues of $\texttt{bcirc}(A)$.

Proof The results follow immediately from the unfolded form of the t-PDS (6.3) and linear stability. $\qquad\square$

The computation of eigenvalues for $\texttt{bcirc}(A)$ can be a computationally expensive task. One can expedite the process by leveraging the Fourier transform.

Proposition 6.4 ([1]) *Given the t-PDS (6.3), the equilibrium point $X_e = 0$ is:*

- *stable if and only if the eigenvalues of A_j have non-positive real parts for all $j = 1, 2, \ldots, r$;*
- *asymptotically stable if and only if the eigenvalues of A_j have negative real parts for all $j = 1, 2, \ldots, r$;*
- *unstable if and only if the eigenvalues of A_j have positive real parts for some $j = 1, 2, \ldots, r$,*

where A_j are the block diagonal matrices from the Fourier transform of A defined in (1.37).

Proof The exponential term $\exp\{\mathtt{bcirc}(\mathsf{A})t\}$ in the solution of the unforced t-PDS (6.3) can be written in the Fourier domain, i.e.,

$$(\mathbf{F}_n \otimes \mathbf{I}_r)\exp\{\mathtt{bcirc}(\mathsf{A})t\}(\mathbf{F}_n^\top \otimes \mathbf{I}_r)$$
$$= \mathtt{blockdiag}\big(\exp\{\mathbf{A}_1 t\}, \exp\{\mathbf{A}_2 t\}, \ldots, \exp\{\mathbf{A}_r t\}\big),$$

where $\mathbf{F}_n \in \mathbb{R}^{n \times n}$ is the Fourier transform matrix. Therefore, the result follows immediately. $\qquad\square$

Remark 6.1 The computational complexity of computing the eigenvalues for $\mathbf{A}_j$ is approximately $\mathcal{O}(n^3 r)$, whereas computing the eigenvalues for $\mathtt{bcirc}\,(\mathsf{A})$ requires approximately $\mathcal{O}(n^3 r^3)$ operations. Interestingly, the eigenvalues of $\mathbf{A}_j$ are intricately linked to the eigentuples of A.

Corollary 6.1 ([1]) *Given the t-PDS (6.3), the equilibrium point $X = 0$ is:*

- *stable if and only if the Fourier transforms of λ_j have non-positive real parts for all $j = 1, 2, \ldots, n$;*
- *asymptotically stable if and only if the Fourier transforms of λ_j have negative real parts for all $j = 1, 2, \ldots, n$;*
- *unstable if and only if the Fourier transforms of λ_j have positive real parts for some $j = 1, 2, \ldots, n$,*

where λ_j are the eigentuples of A.

Proof The result follows immediately from Proposition 6.4 and the fact that the collection of the Fourier transforms of λ_j, i.e., $(\mathbf{F}_n \otimes \mathbf{I}_r)\lambda_j(\mathbf{F}_n^\top \otimes \mathbf{I}_r)$, are the eigenvalues of $\mathtt{bcric}(\mathsf{A})$. $\qquad\square$

6.2.3 Controllability

The controllability of t-PDSs can be formulated similarly as that of LDSs.

Definition 6.2 The t-PDS (6.1) is said to be controllable on the interval $[t_0, t_f]$ if for any two states $X(t_0) = X_0 \in \mathbb{R}^{n \times h \times r}$ and $X(t_f) = X_f \in \mathbb{R}^{n \times h \times r}$ there exists a sequence of control inputs $U(t) \in \mathbb{R}^{s \times h \times r}$ that drive the system from X_0 to X_f.

Proposition 6.5 *The t-PDS (6.1) is controllable if and only if the controllability matrix defined as*

$$\boldsymbol{R} = \begin{bmatrix} \boldsymbol{B} \ \boldsymbol{AB} \ \cdots \ \boldsymbol{A}^{nr-1}\boldsymbol{B} \end{bmatrix} \in \mathbb{R}^{nr \times nsr^2}, \tag{6.6}$$

where $\boldsymbol{A} = \texttt{bcirc}(A)$ and $\boldsymbol{B} = \texttt{bcirc}(B)$, has full rank.

Proof The result follows directly from the matrix form (6.2) of the t-PDS and the Kalman's rank condition. $\square$

Instead of evaluating the rank of the controllability matrix $\mathbf{R}$, one can exploit the singular tuples of the controllability tensor (defined in the following) to determine the controllability.

Corollary 6.2 ([1]) *The t-PDS (6.1) is controllable if and only if all the singular tuples of the controllability tensor defined as*

$$R = \begin{bmatrix} B \ A \star B \ \cdots \ A^{nr-1} \star B \end{bmatrix}_2 \in \mathbb{R}^{n \times nsr \times r} \tag{6.7}$$

contains nonzero entries in the Fourier domain.

Proof The conclusion can be readily derived from Proposition 6.5 and the fact that the rank of the controllability matrix $\mathbf{R}$ is equal to the sum of the nonzero entries in the singular tuples of the controllability tensor R.

Here, the notation $[\cdots]_2$ denotes the block tensor concatenation at the second mode (similar to the definition of p-mode block tensors).

6.2.4 Observability

The observability of t-PDSs can be determined by the dual principal. Suppose that we have an additional output function to the t-PDS (6.1), i.e.,

$$Y(t) = C \star X(t), \tag{6.8}$$

where $C \in \mathbb{R}^{m \times n \times r}$ is the output matrix and $Y(t) \in \mathbb{R}^{m \times h \times r}$. The output function can be expressed in the matrix form as

$$\mathbf{Y}(t) = \texttt{bcirc}(\mathsf{C})\mathbf{X}(t), \tag{6.9}$$

where $\mathbf{Y}(t) = \texttt{unfold}(Y(t))$.

Definition 6.3 The t-PDS (6.1), (6.8) is said to be observable on an interval $[t_0, t_f]$ if any initial state $X(t_0) = X_0 \in \mathbb{R}^{n \times h \times r}$ can be uniquely determined from the outputs $Y(t) \in \mathbb{R}^{m \times h \times r}$.

Proposition 6.6 *The t-PDS (6.1), (6.8) is observable if and only if the observability matrix defined as*

$$\mathbf{O} = \begin{bmatrix} \mathbf{C} \ \mathbf{CA} \ \cdots \ \mathbf{CA}^{nr-1} \end{bmatrix}^{\top} \in \mathbb{R}^{nmr^2 \times nr}, \tag{6.10}$$

where $\mathbf{A} = \texttt{bcirc}(\mathsf{A})$ and $\mathbf{C} = \texttt{bcirc}(\mathsf{C})$, has full rank.

Corollary 6.3 *The t-PDS (6.1), (6.8) is observable if and only if all the singular tuples of the observability tensor defined as*

$$O = \begin{bmatrix} C \ C \star A \ \cdots \ C \star A^{nr-1} \end{bmatrix}_1 \in \mathbb{R}^{nmr \times n \times r} \tag{6.11}$$

contains nonzero entries in the Fourier domain.

Here, the notation $[\cdots]_1$ denotes the block tensor concatenation at the first mode (similar to the definition of p-mode column tensors).

6.3   State Feedback Design

State feedback control is a powerful technique in designing feedback controllers for dynamical systems [13, 14]. It leverages the complete state information of a system to compute the control input. The design of the state feedback gain plays a critical role in ensuring the stability and performance of the closed-loop system. In the context of t-PDSs, it is required to find a state feedback of the form $U(t) = -K \star X(t)$ with $K \in \mathbb{R}^{s \times n \times r}$ such that the closed-loop system

$$\dot{X}(t) = (A - B \star K) \star X(t) \tag{6.12}$$

is (asymptotically) stable.

The design of state feedback for t-PDSs can be efficiently achieved in the Fourier domain, following a similar approach to the computation of t-SVD/eigenvalue decomposition. Assume that the t-PDS system (6.1) is stabilizable, which can be defined similarly using the unfolded form (6.2). The first step involves obtaining the block diagonal matrices

$\mathbf{A}_j$ and $\mathbf{B}_j$ by taking the Fourier transforms of A and B. The next step is to utilize the theory of linear state feedback design to determine $\mathbf{K}_j$ such that the eigenvalues of the matrices $\mathbf{A}_j - \mathbf{B}_j \mathbf{K}_j$ for $j = 1, 2, \ldots, r$ are located in the left-half plane. Finally, the state feedback gain tensor K can be computed by taking the inverse Fourier transform of $\mathbf{K}_j$.

6.4   Applications

6.4.1   Synthetic Data: Stability and Controllability

This example was adapted from [1]. Consider a t-PDS of the form (6.1) with $A \in \mathbb{R}^{2 \times 2 \times 2}$ whose frontal slices are given by

$$A_{::1} = \begin{bmatrix} -6 & 6 \\ -10 & 0 \end{bmatrix} \quad \text{and} \quad A_{::2} = \begin{bmatrix} 0 & 2 \\ 8 & 2 \end{bmatrix},$$

and $B \in \mathbb{R}^{2 \times 1 \times 2}$ whose front slices are given by

$$B_{::1} = B_{::2} = \begin{bmatrix} 1 \\ 1 \end{bmatrix}.$$

The state $X(t) \in \mathbb{R}^{2 \times 1 \times 2}$ with

$$X(t)_{::1} = \begin{bmatrix} x_1(t) \\ x_2(t) \end{bmatrix} \quad \text{and} \quad X(t)_{::2} = \begin{bmatrix} x_3(t) \\ x_4(t) \end{bmatrix}.$$

First, the eigentuples of A in the Fourier domain are computed as

$$\lambda_1 = \begin{bmatrix} -3.414 \\ -0.586 \end{bmatrix} \quad \text{and} \quad \lambda_2 = \begin{bmatrix} -4 + 7.07j \\ -4 - 7.07j \end{bmatrix},$$

where j here represents the imaginary number. Therefore, the open-loop system $\dot{X}(t) = A \star X(t)$ is asymptotically stable, see Fig. 6.1. Second, the controllability tensor $R \in \mathbb{R}^{2 \times 4 \times 2}$ is computed as

$$R_{::1} = \begin{bmatrix} 1 & 2 & -12 & 40 \\ 1 & 0 & -4 & 16 \end{bmatrix} \quad \text{and} \quad R_{::2} = \begin{bmatrix} 1 & 2 & -12 & 40 \\ 1 & 0 & -4 & 16 \end{bmatrix}.$$

The singular tuples of R in the Fourier domain are given by

$$\sigma_1 = \begin{bmatrix} 89.9030 \\ 2.3357 \end{bmatrix} \quad \text{and} \quad \sigma_2 = \begin{bmatrix} 0 \\ 0 \end{bmatrix}.$$

Since one of the singular tuples contains zero entries, the t-PDS is not controllable.

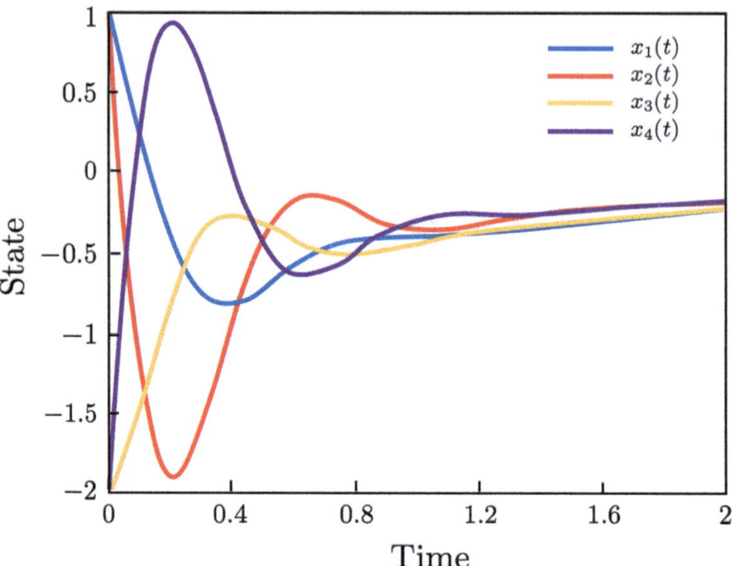

Fig. 6.1 Trajectories of the open-loop t-PDS (i.e., $U(t) = 0$) with initial condition $x_1(0) = x_2(0) = 1$ and $x_3(0) = x_4(0) = -2$. This figure was redrawn from [1] under the Attribution 4.0 International (CC BY 4.0) license

6.4.2 Synthetic Data: State Feedback Design

While the open-loop system of the t-PDS above is already stable, this example, adapted from [1], aims to enhance the closed-loop characteristics by implementing eigenvalues re-assignment in the Fourier domain. First, the block diagonal matrices of A in the Fourier domain are computed as

$$\mathbf{A}_1 = \begin{bmatrix} -6 & 7 \\ -2 & 2 \end{bmatrix} \quad \text{and} \quad \mathbf{A}_2 = \begin{bmatrix} -6 & 3 \\ -18 & -2 \end{bmatrix}.$$

By utilizing the principles of linear state feedback, one can obtain two state feedback gain matrices

$$\mathbf{K}_1 = \begin{bmatrix} 27 & -27 \end{bmatrix} \quad \text{and} \quad \mathbf{K}_2 = \begin{bmatrix} 16.35 & -4.35 \end{bmatrix},$$

which enable the relocation of the eigenvalues of $\mathbf{A}_1$ and $\mathbf{A}_2$ to $-2 \pm 5j$ and $-10 \pm 10j$, respectively. After performing an inverse Fourier transform, the state feedback gain tensor $K \in \mathbb{R}^{1 \times 2 \times 2}$ can be computed as

$$K_{::1} = \begin{bmatrix} 43.35 & -31.35 \end{bmatrix} \quad \text{and} \quad K_{::2} = \begin{bmatrix} 10.64 & -22.64 \end{bmatrix}.$$

By applying the control input $U(t) = -K \star X(t)$, one can achieve the desired closed-loop response, see Fig. 6.2.

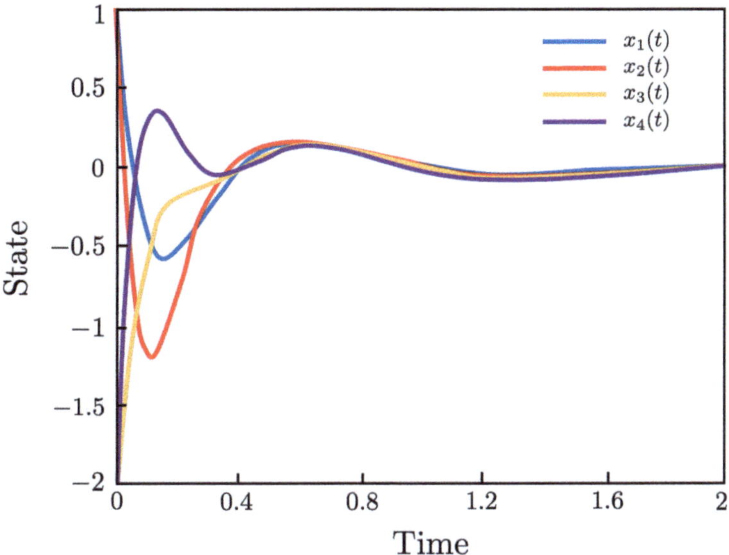

Fig. 6.2 Trajectories of the closed-loop t-PDS (i.e., $U(t) = -K \star X(t)$) with initial condition $x_1(0) = x_2(0) = 1$ and $x_3(0) = x_4(0) = -2$. This figure was redrawn from [1] under the Attribution 4.0 International (CC BY 4.0) license

References

1. Hoover, Randy C., Kyle Caudle, and Karen Braman. 2021. A new approach to multilinear dynamical systems and control. arXiv:2108.13583.
2. Chang, Shih Yu and Yimin Wei. 2022. T-product tensors–part ii: Tail bounds for sums of random t-product tensors. *Computational and Applied Mathematics* 41 (3): 99.
3. Lund, Kathryn. 2020. The tensor t-function: A definition for functions of third-order tensors. *Numerical Linear Algebra with Applications* 27 (3): e2288.
4. Chang, Shih Yu and Yimin Wei. 2021. Generalized t-product tensor bernstein bounds. arXiv:2109.10880.
5. arzanagh, Davoud Ataee, and George Michailidis. 2018. Fast randomized algorithms for t-product based tensor operations and decompositions with applications to imaging data. *SIAM Journal on Imaging Sciences* 11 (4): 2629–2664.
6. Yang, Hang-Jin, Yu-Ying Zhao, Jin-Xing Liu, Yu-Xia Lei, Jun-Liang Shang, and Xiang-Zhen Kong. 2020. Sparse regularization tensor robust pca based on t-product and its application in cancer genomic data. In *2020 IEEE International Conference on Bioinformatics and Biomedicine (BIBM)*, 2131–2138. IEEE.
7. Yu, Na, Zhi-Ping Liu, and Rui Gao. 2022. Predicting multiple types of microrna-disease associations based on tensor factorization and label propagation. *Computers in Biology and Medicine* 146: 105558.

8. Qiao, Qian, Ying-Lian Gao, Sha-Sha Yuan, and Jin-Xing Liu. 2021. Robust tensor method based on correntropy and tensor singular value decomposition for cancer genomics data. In *2021 IEEE International Conference on Bioinformatics and Biomedicine (BIBM)*, 509–514. IEEE.

9. Kilmer, Misha E., Karen Braman, Ning Hao, and Randy C. Hoover. 2013. Third-order tensors as operators on matrices: A theoretical and computational framework with applications in imaging. *SIAM Journal on Matrix Analysis and Applications* 34 (1): 148–172.

10. Kilmer, Misha E., Carla D. Martin, and Lisa Perrone. 2008. A third-order generalization of the matrix svd as a product of third-order tensors. *Tufts University, Department of Computer Science, Tech. Rep. TR-2008-4.*

11. Kilmer, Misha E., Carla D. Martin. 2011. Factorization strategies for third-order tensors. *Linear Algebra and its Applications* 435 (3): 641–658.

12. Zhang, Zemin, and Shuchin Aeron. 2016. Exact tensor completion using t-svd. *IEEE Transactions on Signal Processing* 65 (6): 1511–1526.

13. Kautsky, Jaroslav, Nancy K. Nichols, and Paul Van Dooren. 1985. Robust pole assignment in linear state feedback. *International Journal of control* 41 (5): 1129–1155.

14. Thompson, A.G. 1976. An active suspension with optimal linear state feedback. *Vehicle System Dynamics* 5 (4): 187–203.